SpringerBriefs in Applied Sciences and Technology

SpringerBriefs in Mathematical Methods

Series Editor
Anna Marciniak-Czochra, Institute of Applied Mathematics, IWR, University of Heidelberg, Heidelberg, Germany

Mathematical Methods is a new series of SpringerBriefs devoted to non-standard and fresh mathematical approaches to problems in applied sciences. Compact volumes of 50 to 125 pages, each presenting a concise summary of a mathematical theory, and providing a novel application in natural sciences, humanities or other fields of mathematics. The series is intended for applied scientists and mathematicians searching for innovative mathematical methods to address problems arising in modern research. Examples of such topics include: algebraic topology applied in medical image processing, stochastic semigroups applied in genetics, or measure theory applied in differential equations.

Ryszard Rudnicki • Radosław Wieczorek

Individual-Based Models and Their Limits

Springer

Ryszard Rudnicki
Institute of Mathematics
Polish Academy of Sciences
Katowice, Poland

Radosław Wieczorek
Institute of Mathematics
University of Silesia in Katowice
Katowice, Poland

ISSN 2191-530X ISSN 2191-5318 (electronic)
SpringerBriefs in Applied Sciences and Technology
ISSN 2365-0826 ISSN 2365-0834 (electronic)
SpringerBriefs in Mathematical Methods
ISBN 978-3-031-75269-8 ISBN 978-3-031-75270-4 (eBook)
https://doi.org/10.1007/978-3-031-75270-4

© The Editor(s) (if applicable) and The Author(s), under exclusive license to Springer Nature Switzerland AG 2024

This work is subject to copyright. All rights are solely and exclusively licensed by the Publisher, whether the whole or part of the material is concerned, specifically the rights of translation, reprinting, reuse of illustrations, recitation, broadcasting, reproduction on microfilms or in any other physical way, and transmission or information storage and retrieval, electronic adaptation, computer software, or by similar or dissimilar methodology now known or hereafter developed.
The use of general descriptive names, registered names, trademarks, service marks, etc. in this publication does not imply, even in the absence of a specific statement, that such names are exempt from the relevant protective laws and regulations and therefore free for general use.
The publisher, the authors and the editors are safe to assume that the advice and information in this book are believed to be true and accurate at the date of publication. Neither the publisher nor the authors or the editors give a warranty, expressed or implied, with respect to the material contained herein or for any errors or omissions that may have been made. The publisher remains neutral with regard to jurisdictional claims in published maps and institutional affiliations.

This Springer imprint is published by the registered company Springer Nature Switzerland AG
The registered company address is: Gewerbestrasse 11, 6330 Cham, Switzerland

If disposing of this product, please recycle the paper.

Preface

Individual-based models (IBM) describe a population as a collection of different organisms whose local interactions determine the behavior of the entire population. The individual description is convenient for computer simulations and the determination of various model parameters, and appropriate limit passages lead to the transport equations used in classical population dynamics models. The aim of this book is to provide a brief mathematical introduction to IBMs and their application to selected biological topics.

The book is divided into seven chapters. In the first chapter we give a general description of IBMs and we present examples of models to illustrate their possible applications. Examples of applications include age, size and phenotype models, coagulation-fragmentation process, and models of genome evolution. The second chapter contains some theoretical results concerning limit passages from IBMs to phenotype and age-structured models. The rate of this convergence formulated as functional central limit theorems is presented in Chap. 3. As a result of the limit passage can be a superprocess, i.e., a stochastic process with values in a space of measures. Chapter 4 presents examples of such passages: from the branching Brownian motion to the Dawson–Watanabe superprocess and from the Moran's model of genetic drift with mutations to the Fleming–Viot superprocess.

The next three chapters are devoted to models, in which we directly participated in the study. In Chap. 5 we study IBMs phenotype models and their limit passages. We show that random mating stabilizes the distribution of traits, while assortative mating can lead to a polymorphic population. Formation of aggregates of phytoplankton and their movement is studied in Chap. 6. We present two types of models based on: fragmentation-coagulation processes; and diffusion with chemical signals leading to advanced superprocesses. Chapter 7 is devoted to rather advanced models with chemotaxis used to description of retinal angiogenesis and cell proliferations. In Chap. 7, we also present other selected biological models in which the limit passages from individual models to nonlinear transport equations or superprocesses are investigated. We add to the book Electronic Supplementary Material consisting of two appendices A and B in which we have collected information about stochastic processes and various spaces we have used. In several places in the book we refer to the appendices using the notation such as Section A.2, formula (B.7).

The book is dedicated both to mathematicians and biologists. The first group will find here new biological models which lead to interesting and often new mathematical questions. Biologists can observe how to include seemingly different biological processes into a unified mathematical theory and deduce from this theory interesting biological conclusions. Apart from the sections on superprocesses, where quite advanced mathematical issues arise, such as stochastic partial differential equations, we try to keep the required mathematical and biological background to a minimum so that the topics are accessible to students.

Acknowledgments

This research was partially supported by the National Science Centre (Poland) Grant No. 2017/27/B/ST1/00100. We are grateful to our collaborators Vincenzo Capasso, Katarzyna Pichór, Marta Tyran-Kamińska, and Paweł Zwoleński, with whom we conducted joint research and discussions on many of the issues presented in the monograph.

Katowice, Poland Ryszard Rudnicki
July 2024 Radosław Wieczorek

Contents

1 Models 1
   1.1 Introduction 1
   1.2 General Description of IBMs 2
   1.3 Examples 7
      1.3.1 Size-Structured Population Model 7
      1.3.2 Age-Structured Population Model 8
      1.3.3 Asexual Phenotype Population Model 9
      1.3.4 Phenotype Model with Sexual Reproduction 10
      1.3.5 Coagulation-Fragmentation Processes
           in a Phytoplankton Model 12
      1.3.6 Models of Genome Evolution 13

2 Limit Passages 17
   2.1 From IBMs to Macroscopic Models: Introduction 17
   2.2 From a Jump Process to Diffusion 19
   2.3 Examples of Limit Passages 21

3 Central Limit-Type Theorems 33
   3.1 An Introductory Example 33
   3.2 Poisson Random Measures, Jump Processes and Itô Formula 38
   3.3 Convergence of Càdlàg Processes 40
   3.4 CLT-Type Theorems for IBM 41
   3.5 Examples of CLTs 50

4 Selected Superprocesses 53
   4.1 The Dawson–Watanabe Superprocess 53
   4.2 The Dawson–Watanabe Superprocess as a Solution of a SPDE 57
   4.3 Notes on Other Superprocesses 61

5 Phenotype Models 65
   5.1 From Microscopic to Macroscopic Models 65
   5.2 CLT for a Phenotype Model 70

	5.3 Asymptotic Properties of a Model with Random Mating	72
	5.4 Notes on Other Phenotypic Models	78
6	**Modelling of Phytoplankton Dynamics**	85
	6.1 Model of Growth of Phytoplankton Aggregates	85
	6.2 CLT for Model of Growth of Phytoplankton Aggregates	89
	6.3 Models of Growth and Movement of Aggregates	92
	6.4 Phytoplankton Dynamics as a Superprocess	94
	6.5 Model with Interactions	96
7	**Chemotaxis Models**	99
	7.1 Model of Retinal Angiogenesis	99
	7.2 Branching Diffusion with Location Dependent Branching Rates	103
	7.3 Model of Proliferating Cells with Chemotaxis	111
	7.4 Other IBMs	115
References		119
Index		125

Chapter 1
Models

Individual-based models play a special role in applications of mathematics to biological problems. They describe a population as a heterogeneous family of individuals and the relationships between them. This description is relatively simple and convenient to simulate by computer and to determine various model parameters. Moreover, a macroscopic description of the population can be obtained from an individual-based model by appropriate limits. In this chapter we give a general characterization of individual-based models and we present examples of models to illustrate their possible applications. Examples of applications include age, size and phenotype models, coagulation-fragmentation process, and models of genome evolution.

1.1 Introduction

In general, a population does not have a homogeneous structure. Individuals differ in characteristics such as age, maturity in the case of cells, phenotype, location in space or group membership. Models that take population structure into account are called *structured models*. Among structured models, age and phenotypic models are distinguished, and in adaptive dynamics the features that distinguish individuals are called *strategies*. Another example of a structured model is one that describes phytoplankton aggregates or other cell colonies, and in this case individuals are aggregates/colonies, and they differ in size.

Models in which each individual has individual characteristics and we are interested in interactions between individuals and their effect on the population as a whole are called *individual models*. Individual models are referred to by the terms: *individual-based models* (IBMs) [35, 52, 75, 104] or *agent-based models* (ABMs) [4, 14, 88]. The state of a population in an IBM is the set of characteristics of all individuals and we are interested in how this set changes over time.

Individual-based models are also called *microscopic models*, or *lagrangian*, because they describe dynamics at the level of individuals, as opposed to *macroscopic models*, which describe changes in the state of the entire population. Some authors [71] additionally distinguish intermediate models called *mesoscopic models*.

It is quite natural to describe biological issues, especially in ecology, using models based on the behaviour of individuals [35, 52, 75, 79, 104] and this description has a number of advantages. It allows us to define the relationships between individuals in a population in a relatively easy way, and the various elements of the model: death of individuals, mating, offspring, competition, etc. are introduced in a transparent way. Secondly, the individual description is simple and convenient to simulate by computer and to determine various model parameters. It is more challenging to perform simulations of transport equations traditionally used in the *macroscopic description* of the evolution of distributions. In individual models it is easy to follow what happens to related individuals. Thirdly, a macroscopic description, also called an *eulerian description* can be obtained from the individual description by appropriate limits [37, 45, 70, 82]. Then transport equations appear naturally, rather than being introduced *ad hoc*.

Although the description of biological problems by means of IBMs is relatively simple, the study of passage from IBMs to transport equations or to stochastic processes describing the behaviour of the whole population is an issue requiring the use of advanced mathematical methods. The aim of the book is to give a general description of models that are both structured and individual and to present examples of applications and results on their behaviour. In particular we will be interested in limit passages from IBMs to transport equations and to superprocesses, i.e. stochastic processes with values in the space of measures.

1.2 General Description of IBMs

We consider a finite population in which each individual is characterized by a certain *trait*. This trait can change over time. Let $\mathscr{X}$ be the set of all possible traits. The set $\mathscr{X}$ can be arbitrary, but usually $\mathscr{X}$ is a subset of $\mathbb{R}^d$ or as in the next few sections it will be a countable set (finite or infinite). At time t, the population is composed of $k(t)$ individuals with features $x_1(t), x_2(t), \ldots, x_k(t)$. We can assume that the state of the population at time t is the set

$$\{x_1(t), \ldots, x_k(t)\},$$

and its evolution is described by a stochastic process $\xi_t = \{x_1(t), \ldots, x_k(t)\}$. Since the values of the process are sets of points from the set $\mathscr{X}$, the process $(\xi_t)_{t \geq 0}$ is called a *point process*. This approach is quite natural, but has some drawback—individuals with the same characteristics occur only once in the considered set.

1.2 General Description of IBMs

We can modify this description and assume that $(\xi_t)_{t\geq 0}$ is a process whose values are multisets. A *multiset* (or a *bag*) is a generalization of the concept of a set in which, unlike classical sets, a single element can occur multiple times.

Another solution of this problem is to consider $(\xi_t)_{t\geq 0}$ as a process with values in the space of measures. In our case

$$\xi_t = \delta_{x_1(t)} + \cdots + \delta_{x_{k(t)}(t)}, \qquad (1.1)$$

where δ_a denotes the *Dirac measure* at the point a, i.e., δ_a is the probability measure concentrated at the point a. Since ξ_t is a counting measure, $\xi_t(A)$ is the number of individuals with traits from the set A at time t. Also this approach has some disadvantages, for example it is rather difficult to consider differential equations on measures.

We can also consider a state of the system as a k-tuple $(x_1(t), \ldots, x_k(t))$. With this approach, another problem arises. The size of the population may vary, and it is quite inconvenient to operate with strings of different lengths. To avoid this problem, a "death state" $*$ is introduced, and then the population state is an infinite sequence of elements from the space $\mathscr{X}_* = \mathscr{X} \cup \{*\}$, which has points $x_1(t), \ldots, x_k(t)$ on some places, and on the remaining $*$. Let us denote by X the space of all strings of which at most a finite number have values in $\mathscr{X}$ and the others have $*$. The space X is not yet a proper state space of the population, because such a description is not explicit. In order to have uniqueness in the space X we introduce an equivalence relation $\sim$. The two strings $\mathbf{x} = (x_1, x_2, \ldots)$ and $\mathbf{y} = (y_1, y_2, \ldots)$ are equivalent if $\mathbf{y}$ is a permutation of $\mathbf{x}$, i.e. $\mathbf{x} \sim \mathbf{y}$ if and only if there exists a bijection $\sigma: \mathbb{N}_+ \to \mathbb{N}_+$ such that $\mathbf{y} = (x_{\sigma(1)}, x_{\sigma(2)}, \ldots)$. We assume that the state space $\widetilde{X}$ in our model is the set of equivalence classes with respect to $\sim$, so $\widetilde{X} = X/\sim$.

The question arises how to define a σ-algebra Σ of subset of $\widetilde{X}$? Usually, Σ is the σ-algebra of Borel subsets of $\widetilde{X}$, thus we need to introduce a topology on the space S. We can identify the space $\widetilde{X}$ with the space $\mathscr{N}$ of finite counting measures on $\mathscr{X}$ by a map $\eta: \widetilde{X} \to \mathscr{N}$ given by

$$\eta(\mathbf{x}) = \sum_{\{i:\, x_i \neq *\}} \delta_{x_i} \qquad (1.2)$$

where $\mathbf{x} = [(x_1, x_2, \ldots)]$. One of the measures in $\mathscr{N}$ is the null measure $\mathfrak{O}$ which is identified with the sequence $[(*, *, \ldots)]$. The space $\mathscr{N}$ is equipped with the topology of weak convergence of measures, so we define open sets in $\widetilde{X}$ as preimages through the function η of open sets in $\mathscr{N}$. Thus, the population is described by a process $\xi_t = [(x_1(t), x_2(t), \ldots)]$ with values in $\widetilde{X}$ or equivalently by the process $\bar{\xi}_t = \eta(\xi_t)$, $t \geq 0$, with values in $\mathscr{N}$.

Assuming that ρ is a metric on the space $\mathscr{X}_*$, we can also define a metric on the space $\widetilde{X}$. First, we define a metric d on the space X by

$$d(\mathbf{x}, \mathbf{y}) = \max_{i \in \mathbb{N}} \rho(x_i, y_i)$$

and then we define a metric $\tilde{d}$ on the space $\tilde{X}$ by

$$\tilde{d}([\mathbf{x}], [\mathbf{y}]) = \min\{d(\mathbf{a}, \mathbf{b}) \colon \mathbf{a} \in [\mathbf{x}], \mathbf{b} \in [\mathbf{y}]\}.$$

In the description of an individual model, there are features that can change at any time moment, such as an individual's position in space and size, and traits that change in jumps, for example when moving to another group of individuals, or are constant throughout the life of an individual (such as sex, phenotype). In addition to jump changes, we add births and deaths of individuals.

Till now we have only described the state space of the process $(\xi_t)_{t\geq 0}$. We restrict our investigations to time homogeneous Markov process on the state space $\tilde{X}$, but a general description of IBMs is still complicated, so we will do it in some steps starting with case when $\mathscr{X}$ is a countable set. Then $\tilde{X}$ is also a countable set. We assume that the *transition probability* from a state $\mathbf{x}$ to a state $\mathbf{y} \neq \mathbf{x}$ in time Δt:

$$p_{\mathbf{xy}}(\Delta t) = \mathrm{P}(\xi_{t+\Delta t} = \mathbf{y} | \xi_t = \mathbf{x})$$

does not depend on t. We also assume that the limit

$$q_{\mathbf{xy}} = \lim_{\Delta t \to 0} \frac{p_{\mathbf{x},\mathbf{y}}(\Delta t)}{\Delta t} \qquad (1.3)$$

exists. This limit is called *transition rate* from $\mathbf{x}$ to $\mathbf{y}$. We also assume that the *jump rate function* $\varphi(\mathbf{x})$ given by

$$\varphi(\mathbf{x}) = \sum_{\mathbf{y} \neq \mathbf{x}} q_{\mathbf{xy}}$$

is finite for each $\mathbf{x} \in \tilde{X}$ and define $q_{\mathbf{xx}} = -\varphi(\mathbf{x})$. The matrix $Q = [q_{\mathbf{xy}}]$ is called the *transition rate matrix* or the *infinitesimal generator matrix*. The matrix $Q = [q_{\mathbf{xy}}]$ has the following properties

(i) $q_{\mathbf{xy}} \geq 0$ for $\mathbf{x} \neq \mathbf{y}$,
(ii) $\sum_{\mathbf{y} \in \tilde{X}} q_{\mathbf{xy}} = 0$ or each $\mathbf{x} \in \tilde{X}$.

A matrix $Q = [q_{\mathbf{xy}}]$ having properties (i)–(ii) is called a *Kolmogorov matrix*. The matrix Q plays a crucial role in the theory of Markov processes. In particular, if $p_{\mathbf{x}}(t) = \mathrm{P}(\xi_t = \mathbf{x})$, then the vector $p(t)$ with coordinates $p_{\mathbf{x}}(t)$ satisfies the following system of differential equations

$$p'(t) = p(t)Q.$$

1.2 General Description of IBMs

In applications we usually know a Kolmogorov matrix Q, and the question is if Q is a transition rate matrix of some Markov chain? The problem is non-trivial and the following theorem answers this question.

Theorem 1.1 (Kato) *Let Q be a Kolmogorov matrix and let $\lambda > 0$ be a positive constant. Then Q is a transition rate matrix of some Markov chain if and only if there is no non-trivial bounded sequence w such that $Qw = \lambda w$.*

Thus, if the matrix Q with entries defined by (1.3) satisfies assumptions of Theorem 1.1, then $(\xi_t)_{t \geq 0}$ is a continuous-time Markov chain with values in the space $\widetilde{X}$. A Kolmogorov matrix Q having the above property is called *non-explosive*. In particular if $|q_{\mathbf{xx}}| \leq C$ for some $C > 0$ and all $\mathbf{x} \in \widetilde{X}$, then the matrix Q is non-explosive.

Markov chains with continuous time belong to a somewhat broader class of *pure-jump Markov processes*. The space $\mathscr{X}$ can be any measurable space. Pure-jump Markov processes are used to describe *phenotype-structured population models*. Such models are described by two factors: a measurable *jump rate function* $\varphi \colon \widetilde{X} \to [0, \infty)$ and a *transition probability function* $P(\mathbf{x}, A)$, on $\widetilde{X}$, i.e., $P(\mathbf{x}, \cdot)$ is a probability measure for each $\mathbf{x} \in \widetilde{X}$ and the function $\mathbf{x} \mapsto P(\mathbf{x}, B)$ is measurable for each $B \in \Sigma$. Let $t_0 = 0$ and let ξ_0 be an $\widetilde{X}$-valued random variable. For each $n \geq 1$ one can choose the *n*th *jump time* t_n as a positive random variable satisfying

$$P(t_n - t_{n-1} \leq t | \xi_{t_{n-1}} = \mathbf{x}) = 1 - e^{-\varphi(\mathbf{x})t}, \quad t \geq 0,$$

and define

$$\xi_t = \begin{cases} \xi_{t_{n-1}} & \text{for } t_{n-1} \leq t < t_n, \\ \xi_{t_n} & \text{for } t = t_n, \end{cases}$$

where the *n*th *post-jump position* ξ_{t_n} is an $\widetilde{X}$-valued random variable such that

$$P(\xi_{t_n} \in B | \xi_{t_{n-1}} = \mathbf{x}) = P(\mathbf{x}, B).$$

The process $(\xi_t)_{t \geq 0}$ is well defined if $\lim_{n \to \infty} t_n = \infty$ a.s.; the last condition is fulfilled if, for example, φ is a measurable bounded function.

Pure-jump Markov processes belong to a broader class of *piecewise deterministic Markov processes* (PDMPs). The concept of PDMPs was introduced by Davis [32]. According to his informal definition, PDMPs are a general family of stochastic models covering virtually all non-diffusive applications of Markov processes. However, such an outlined definition is nowadays too broad and needs to be clarified. At this point we only give a descriptive definition, without going into mathematical details. A Markov process $(\xi_t)_{t \geq 0}$ with values in some metric space X is called a PDMP if there exists an increasing sequence of random times (t_n), called *jump moments*, such that the sample paths (trajectories) of the process are defined in a deterministic way in each interval (t_n, t_{n+1}). We distinguish two types of process behaviour at jump moments: a sample path of the process can jump to a random place in space or we can change dynamics determining the sample paths. Jump moments can depend on the point through which the trajectory passes. PDMPs form a family of various stochastic processes including a lot of different biological models [91]. Despite their great diversity, PDMPs stand out among other Markov processes one important feature. They are usually defined in a relatively simple and constructive way.

The last group of IBMs considered by us are models with stochastic equations. Let $0 < t_1 < t_2 < \ldots$ be random moments of jump changes, which also include moments of birth and death of individuals in the population. We assume that in the interval $[t_{n-1}, t_n)$ the dynamics of trait change is described by the following system of stochastic equations

$$dX_t^i = b_{k(t)}(\xi_t, X_t^i)\,dt + \sum_{j=1}^{k(t)} \sigma_{k(t)}^j(\xi_t, X_t^i)\,dW_t^j, \quad i = 1, \ldots, k(t), \qquad (1.4)$$

where X_t^i describes the characteristics of i-th individual changing continuously, $k(t)$ is the number of individuals in the population, $W_t^1, \ldots, W_t^{k(t)}$ are independent d-dimensional Wiener processes, and b_k and σ_k^j are functions defined on the set $\mathcal{N} \times \mathcal{X}$ with values in $\mathbb{R}^d$ and $\mathbb{M}^{d \times d}$, respectively. Here $\mathbb{M}^{d \times d}$ denotes the set of real matrices with d rows and d columns. The process $(\xi_t)_{t \geq 0}$ is given by the formula (1.1) and describes the state of the entire population. We see that in the i-th equation the right-hand side of the equation depends only on the state of the whole population and the characteristics of the i-th individual. At first glance, the system of (1.4) seems complicated, but it is still a system of stochastic equations, because in the whole interval $[t_{n-1}, t_n)$ the functions occurring on the right side of the system depend only on variables on the left-hand side of the system. At times of jump t_n, both the population size and the characteristics of individuals can change. If $\mu = \xi_{t_n^-}$, then ξ_{t_n} is a measure drawn from the distribution of $\mathscr{P}(\mu, \cdot)$, where $\mathscr{P} \colon \mathcal{N} \times \mathscr{B}(\mathcal{N}) \to [0, 1]$ is the transition function and $\mathscr{B}(\mathcal{N})$ is the σ-algebra of Borel subsets of $\mathcal{N}$ in the topology of weak convergence of measures. The jump moments (t_i) are chosen with some intensity φ dependent on the state of the population ξ_t. This procedure allows us to define a process $(\xi_t)_{t \geq 0}$ having its initial value ξ_0 fixed. The process $(\xi_t)_{t \geq 0}$ is a Markov process with values in the measure space $\mathcal{N}$.

Note that the individual models obtained by piecewise deterministic Markov processes come under the above scheme. In this case, the system of equations (1.4) is deterministic, so it has no components of the form $\sum_{j=1}^{k(t)} \sigma_{k(t)}^j(\xi_t, X_t^i)\,dW_t^j$. Examples are the age-structure and coagulation-fragmentation models from Sect. 1.3.

1.3 Examples

1.3.1 Size-Structured Population Model

We now present an individual structured model of a cell population in which cells are characterized by their size. In this case $\mathscr{X} = (0, \infty)$. We assume that the size $x(t)$ of a cell grows according to the equation

$$x'(t) = g(x(t)).$$

A single cell with size x replicates with rate $b(x)$ and dies with rate $d(x)$. A daughter cell has a half size of the mother cell. The evolution of the population is a Markov process $\xi_t = [(x_1(t), x_2(t), \dots)]$, $t \geq 0$, with values in the space $\widetilde{X}$. The process $(\xi_t)_{t \geq 0}$ has jump points when one of the cells dies or replicates. We define $g(*) = b(*) = d(*) = 0$ and assume that $x(t) = *$ is the solution of the equation $x'(t) = 0$ with the initial condition $x(0) = *$. Between jumps trajectories satisfy the equation

$$[(x_1'(t) - g(x_1(t)), x_2'(t) - g(x_2(t)), \dots)] = [(0, 0, \dots)]. \tag{1.5}$$

For $t \geq 0$ and $x^0 \in \mathbb{R}_*^+$ we denote by $\pi(t, x^0)$ the solution $x(t)$ of the equation $x'(t) = g(x(t))$ with the initial condition $x(0) = x^0$. Let $\mathbf{x}^0 = [(x_1^0, x_2^0, \dots)] \in \widetilde{X}$ and define

$$\widetilde{\pi}(t, \mathbf{x}^0) = [(\pi(t, x_1^0), \pi(t, x_2^0), \dots)].$$

The jump rate function $\varphi(\mathbf{x})$ at the state $\mathbf{x} = [(x_1, x_2, \dots)]$ is the sum of rates of death and division of all cells:

$$\varphi(\mathbf{x}) = \sum_{i=1}^{\infty} (b(x_i) + d(x_i)). \tag{1.6}$$

If $\mathbf{x}^0 \in \widetilde{X}$ is the initial state of the population at a jump time t_n, then the probability distribution function of $t_{n+1} - t_n$ is given by

$$F(t) = 1 - \exp\left\{-\int_0^t \varphi(\widetilde{\pi}(s, \mathbf{x}^0))\, ds\right\}. \tag{1.7}$$

At time t_n one of the cells dies or replicates. If a cell dies we change the sequence by removing its size from the sequence and we have

$$\mathrm{P}\big(\xi_{t_n} = [(x_1(t_n^-), \dots, x_{i-1}(t_n^-), x_{i+1}(t_n^-), x_{i+2}(t_n^-), \dots)]\big) = \frac{d(x_i(t_n^-))}{\varphi(\xi(t_n^-))}$$

for $i \in \mathbb{N}_+$. If a cell replicates we remove its size from the sequence and add two new elements in the sequence with the sizes of the daughter cells and we have

$$P\big(\xi_{t_n} = [(x_1(t_n^-), \ldots, x_{i-1}(t_n^-), \tfrac{1}{2}x_i(t_n^-), \tfrac{1}{2}x_i(t_n^-), x_{i+1}(t_n^-), x_{i+2}(t_n^-), \ldots)]\big)$$
$$= \frac{b(x_i(t_n^-))}{\varphi(\xi(t_n^-))}$$

for $i \in \mathbb{N}_+$. In this way we have checked that the point process $\xi(t)$, $t \geq 0$, is a homogeneous PDMP with values in $\widetilde{X}$.

1.3.2 Age-Structured Population Model

We consider a population of one sex (usually females) in which all individuals are characterized by their age $x \geq 0$. We assume that an individual with age x dies with rate $d(x)$ and gives birth to one child with rate $b(x)p_1$ or twins with rate $b(x)p_2$, where $p_1 + p_2 = 1$. The model is very similar to the size-structured population. We assume that $g(*) = b(*) = d(*) = 0$ and $g(x) = 1$ for $x \in \mathbb{R}_+$. Then the evolution of the population is described by a stochastic process $\xi_t = [(x_1(t), x_2(t), \ldots)]$, $t \geq 0$, with values in the space $\widetilde{X}$ with jumps when an individual dies or gives birth. The jump rate function $\varphi(\mathbf{x})$ is given by (1.6) and between jumps trajectories fulfills Eq. (1.5). If t_n is a moment of jump and $\xi_{t_n^-} = \mathbf{x} = [(x_1, x_2, \ldots)]$, then ξ_{t_n} takes one of the following values:

$[(x_1, \ldots, x_{i-1}, x_{i+1}, \ldots)]$ with probability $d(x_i)/\varphi(\mathbf{x})$ for $i \in \mathbb{N}_+$,

$[(0, x_1, x_2, \ldots)]$ with probability $b(\mathbf{x})p_1/\varphi(\mathbf{x})$, (1.8)

$[(0, 0, x_1, x_2, \ldots)]$ with probability $b(\mathbf{x})p_2/\varphi(\mathbf{x})$,

where $b(\mathbf{x}) = \sum_{i=1}^{\infty} b(x_i)$.

A more general model can be considered, when the birth and death rates depend not only on the age of an individual but also on the state of the whole population $\mathbf{x} = [(x_1, x_2, \ldots)]$, e.g. on the size of the population. If $b(x_i, \mathbf{x})$, $d(x_i, \mathbf{x})$ are the birth and death rates, respectively, for an individual with age x_i, then the jump rate function is given by

$$\varphi(\mathbf{x}) = \sum_{i=1}^{\infty} (b(x_i, \mathbf{x}) + d(x_i, \mathbf{x})) \quad (1.9)$$

and replace $b(x_i)$ and $d(x_i)$ by $b(x_i, \mathbf{x})$ and $d(x_i, \mathbf{x})$ in the formulas (1.8). Also in this case $(\xi_t)_{t \geq 0}$ is a homogeneous PDMP.

1.3 Examples

1.3.3 Asexual Phenotype Population Model

We further consider a population composed of one sex. Each individual is characterised by its phenotype a vector x from a set $\mathscr{X} \subset \mathbb{R}^d$. We assume that the phenotype does not change over the lifetime of the individual. Substituting age for phenotype, we introduce the coefficients $d(x_i, \mathbf{x})$ and $b(x_i, \mathbf{x})$ as in the previous model. Since the phenotype of progeny may differ from that of the parent, so we assume that if the parent had a phenotype x, then the child has a phenotype chosen according to the transition function $P(x, B)$. Here we consider a simplified model in which the phenotype of the twins are identical (but it is not difficult to build a model without this assumption). We obtain a pure-jump homogeneous Markov process. The jump rate function $\varphi(\mathbf{x})$ is given by (1.9). If $\xi_{t_n^-} = \mathbf{x} = [(x_1, x_2, \ldots)]$, then

$$P(\xi_{t_n} = [(x_1, \ldots, x_{i-1}, x_{i+1}, \ldots)]) = \frac{d(x_i, \mathbf{x})}{\varphi(\mathbf{x})},$$

$$P(\xi_{t_n} = [(x, x_1, x_2, \ldots)]: x \in B) = \sum_{i=1}^{\infty} \frac{p_1 b(x_i, \mathbf{x}) P(x_i, B)}{\varphi(\mathbf{x})},$$

$$P(\xi_{t_n} = [(x, x, x_1, x_2, \ldots)]: x \in B) = \sum_{i=1}^{\infty} \frac{p_2 b(x_i, \mathbf{x}) P(x_i, B)}{\varphi(\mathbf{x})}.$$

It should be noted that in cellular populations we need to change the last two formulas because progeny consists of two cells ($p_1 = 0$) and we lose a cell which replicates. In this case we have

$$P(\xi_{t_n} = [(x, x, x_1, \ldots, x_{i-1}, x_{i+1}, \ldots)]: x \in B) = \frac{b(x_i, \mathbf{x}) P(x_i, B)}{\varphi(\mathbf{x})}.$$

Also the assumption that both daughter cells have the same phenotype is generally not correct. If we consider hematopoietic stem cells—precursors of blood cells living in the bone marrow—then such cells can be at different levels of morphological development. After division a daughter cell can remain on the same level as the mother cell or go to the next level.

Remark 1.1 Since the space $\widetilde{X}$ is the space of equivalence classes in the relation $\sim$, we can write in our formulae arbitrary element from the same class. For example, at the death of an individual with phenotype x_i, we go from state $\mathbf{x} = [(x_1, x_2, \ldots)]$ to the state $[(x_1, \ldots, x_{i-1}, x_{i+1}, \ldots)]$. For practical reasons related to computer simulations of these processes, it may be more convenient to write that we go to $[(x_1, \ldots, x_{i-1}, *, x_{i+1}, \ldots)]$. Then we still have an element from the same class, but we only make a change at one place in the sequence: we replace x_i with $*$. Similarly, for the birth of a single individual with the phenotype x_i, we only need to replace one $*$ by x_i.

1.3.4 Phenotype Model with Sexual Reproduction

The study of phenotype inheritance has a long tradition and began with the seminal article of Fisher [46]. The reader will find extensive information on classical models of inheritance and evolution of phenotypic traits in the book [21]. Modelling populations with sexual reproduction is a challenging task because such processes depend on many factors, including the social behaviour of individuals. The main problem arises when describing the pairing of individuals [96]. The simplest situation is when each individual has both male and female reproductive organs, which is the case in most plants and invertebrates. Then the description can be restricted to semi-random or assortative (selective) pairing (mating).

In the case of semi-random mating we assume that an individual with phenotype $x \in \mathscr{X}$ has mating rate $p(x)$. It means that if the population consists of k individuals with phenotypes $x_1, x_2, \ldots, x_k$ then individuals with phenotype x_i and x_j form a pair with rate

$$p(x_i, x_j) = \frac{p(x_i)p(x_j)}{\sum_{r=1}^{k} p(x_r)}. \tag{1.10}$$

If we exclude self-fertilization then this rate can be given by

$$p(x_i, x_j) = \frac{1}{2}\frac{p(x_i)p(x_j)}{\sum_{r=1, r\neq i}^{k} p(x_r)} + \frac{1}{2}\frac{p(x_i)p(x_j)}{\sum_{r=1, r\neq j}^{k} p(x_r)} \tag{1.11}$$

for $i \neq j$ and $p(x_i, x_j) = 0$ if $i = j$. We assume that if x_1 and x_2 are the parents' phenotypes, then the progeny phenotype is from a set $B \subset \mathscr{X}$ with probability $P(x_1, x_2, B)$, where $P\colon \mathscr{X} \times \mathscr{X} \times \mathscr{B}(\mathscr{X}) \to [0, 1]$ is a transition probability from $\mathscr{X} \times \mathscr{X}$ to $\mathscr{X}$, i.e., $P(x_1, x_2, \cdot)$ is a probability measure on $\mathscr{B}(\mathscr{X})$ for all x_1, x_2 and for each Borel set $B \subset \mathscr{X}$, $(x_1, x_2) \mapsto P(x_1, x_2, B)$ is a measurable function. For simplicity we consider a model when the result of reproduction is only one child. If $\mathbf{x} = [(x_1, x_2, \ldots)]$ is the state of the population then $d(x_i, \mathbf{x})$ is the death rate for an individual with phenotype x_i. As in the previous model the population is described by a pure-jump homogeneous Markov process $\xi_t = [(x_1(t), x_2(t), \ldots)]$. The jump rate function $\varphi(\mathbf{x})$ is given by

$$\varphi(\mathbf{x}) = \sum_{i=1}^{\infty} d(x_i, \mathbf{x}) + \sum_{i=1}^{\infty}\sum_{j=1}^{\infty} p(x_i, x_j), \tag{1.12}$$

1.3 Examples

where $d(*, \mathbf{x}) = 0$ and $p(*, x_i) = p(x_i, *) = 0$. If $\xi_{t_n^-} = \mathbf{x} = [(x_1, x_2, \ldots)]$, then

$$P(\xi_{t_n} = [(x_1, \ldots, x_{i-1}, x_{i+1}, \ldots)]) = \frac{d(x_i, \mathbf{x})}{\varphi(\mathbf{x})},$$

$$P(\xi_{t_n} = [(x, x_1, x_2, \ldots)] : x \in B) = \sum_{i=1}^{\infty} \sum_{j=1}^{\infty} \frac{p(x_i, x_j) P(x_i, x_j, B)}{\varphi(\mathbf{x})}.$$

In the case of *assortative mating* individuals with similar traits mate more often than if they were choosing a partner at random. Assortative mating can be modelled by matching theory, according to which each individual ranks all potential partners according to its preferences and attempts to pair with the one with highest-ranking [3], or by using the preference function [50, 95]. In assortative mating a preference function $a(x, y)$ is usually of the form $a(x, y) = \psi(\|x - y\|)$, where $\psi: [0, \infty) \to [0, \infty)$ is a continuous and decreasing function. We assume that two individuals with phenotypes x_i and x_j form a pair with rate

$$p(x_i, x_j, \mathbf{x}) = \frac{a(x_i, x_j)}{\sum_{l=1}^{\infty} a(x_i, x_l)} = \frac{\psi(\|x_i - x_j\|)}{\sum_{l=1}^{\infty} \psi(\|x_i - x_l\|)} \quad (1.13)$$

and the rest of the model is similar to that with semi-random mating.

If we consider a two-sex population (i.e. each individual is exclusively male or female), then the mating process is more complex and there are only a few mathematical models of it (see e.g. [69] for insect populations). Since the role of males and females is different, it is clear that in such models we should characterize an individual by its sex and phenotype. An important factor is a sexual selection [64] which often depends on the social behavior of individuals. Males usually produce enough sperm to fertilize many females, and when they do not participate in parental care, their reproductive success depends on their phenotype. We give some hints how to build such a model. First, each individual is described by a pair (x, s), where x is its phenotype, $s = 0$ if it is a male and $s = 1$ if it is a female. Let $\mathbf{x} = [((x_1, s_1), (x_2, s_2), \ldots)]$ be the state of the population. We assume that a female with phenotype x_i gives birth with rate $b(x_i, \mathbf{x})$ and a male with phenotype x_j has the competition rate $p(x_j, \mathbf{x})$. To simplify notation we set $b(x_i, \mathbf{x}) = 0$ if $s_i = 0$ and $p(x_j, \mathbf{x}) = 0$ if $s_i = 1$. A female with phenotype x_i and a male with phenotype x_j form a pair with rate

$$p(x_i, x_j, \mathbf{x}) = \frac{b(x_i, \mathbf{x}) p(x_j, \mathbf{x})}{\sum_{r=1}^{\infty} p(x_r, \mathbf{x})}. \quad (1.14)$$

The rest of the model is similar to the hermaphroditic population model and we omit it here.

1.3.5 Coagulation-Fragmentation Processes in a Phytoplankton Model

Mathematical modelling of plankton behaviour is a complex issue involving various mathematical tools. A review of mathematical models of plankton dynamics can be found in [94]. One of the issues being investigated is the formation of plankton aggregates in which they live together like colonial organisms. The size of aggregates is important because they are food in the initial phase of fish development and therefore modelling the aggregate size distribution is an interesting biological and mathematical problem. We now present a relatively simple model of the structure of aggregates [5], in which we consider their growth, extinction and coagulation and fragmentation processes. A more advanced model including their spatial distribution and diffusion [93] will be presented in Sect. 6.3.

In this model individuals are aggregates characterized by their size $x > 0$, which depends on the number of cells. The division or death of individual cells changes the size of aggregates according to the equation $x'(t) = g(x(t))$ but aggregates can die, for example, by sinking to a seabed or whatever cause, with mortality rate $d(x)$. An aggregate can break into two aggregates with intensity $b(x)$ and the size of the aggregates after the breakup is given by the conditional density $k(y, x)$, where k satisfies $k(y, x) = k(x - y, x)$. We have also a coagulation (aggregation) process, by which two distinct aggregates join together to form a single one. We assume that the ability to glue to another aggregate is given by the function $p(x)$. Since the same aggregate cannot join with itself, we assume that two aggregates with sizes x_i and x_j form a new aggregate with rate $p(x_i, x_j)$ given by (1.11) for $i \neq j$ and $p(x_i, x_j) = 0$ if $i = j$.

Our model is similar to that from Sect. 1.3.1. The population of aggregates is described by a homogeneous Markov process $(\xi_t)_{t \geq 0}$ with values in $\widetilde{X}$ and satisfies Eq. (1.5). It remains to incorporate to the model fragmentation and coagulation processes. We have a jump if one of aggregates dies, splits or two aggregates form a new one. The jump rate function $\varphi(\mathbf{x})$ at state $\mathbf{x} = [(x_1, x_2, \ldots)]$ is the sum of mortality, fragmentation and coagulation rates:

$$\varphi(\mathbf{x}) = \sum_{i=1}^{\infty}(d(x_i) + b(x_i)) + \sum_{i=1}^{\infty}\sum_{j=1}^{\infty} p(x_i, x_j), \qquad (1.15)$$

where $d(*) = b(*) = p(x_i, *) = p(*, x_i) = 0$. If $\xi_{t_n^-} = \mathbf{x} = [(x_1, x_2, \ldots)]$, then

$$P(\xi_{t_n} = [(x_1, \ldots, x_{i-1}, x_{i+1}, \ldots)]) = \frac{d(x_i)}{\varphi(\mathbf{x})},$$

$$P(\xi_{t_n} = [(x_1, \ldots, x_{i-1}, y, x_i - y, x_{i+1}, \ldots)]: y \in B) = \frac{b(x_i)\int_B k(y, x_i)\, dy}{\varphi(\mathbf{x})},$$

$$P(\xi_{t_n} = [(x_i + x_j, x_1, \ldots, x_{i-1}, x_{i+1}, \ldots, x_{j-1}, x_{j+1}, \ldots)]) = \frac{p(x_i, x_j)}{\varphi(\mathbf{x})}.$$

1.3.6 Models of Genome Evolution

In [97] it is investigated a model for the evolution of the distribution of paralog families in the genome. We briefly present the model but our aim is to show that the same issue can be modelled by different Markov chains.

In each genome some genes occur several times. These are genes derived from a common ancestor, resulting from a duplication of the same gene. We are not able to distinguish between such genes; we consider them to be identical and we call them *paralogs*.

The model from [97] is based on three basic evolutionary events: gene duplication, deletion (or inactivation) and accumulated change, called for simplicity mutation. We assume that genes resulting from mutations were not previously present in the genome. A single gene:

- *duplicates* with rate d,
- *is removed* from the genome with rate r,
- *mutates* with rate m and then the gene starts a new one-element paralog family (a new gene).

Moreover, we assume that all elementary events are independent of each other. As a result of fundamental evolutionary events a family of paralogs of size x:

- increases to size $x + 1$ with rate dx,
- decreases to size $x - 1$ with rate rx,
- decreases to size $x - 1$, and forms a new family of size 1 with rate mx.

Now an individual is a paralog family and the trait is the size of the family, so $\mathscr{X} = \mathbb{N}_+$. If a family has only one gene and this gene is remove, then we replace this family by $* = 0$. Thus $\mathscr{X}_* = \mathbb{N}$ and X is the space of all sequences with values in $\mathbb{N}$ different from zero for at most a finite number of terms. The model is described by a homogeneous Markov process $(\xi_t)_{t \geq 0}$ on the space $\widetilde{X} = X/\sim$. The intensity matrix Q has the following entries different from zero:

$$q_{[(x_1,x_2,\dots)],[(x_1,\dots,x_{n-1},x_n+1,x_{n+1},\dots)]} = dx_n,$$

$$q_{[(x_1,x_2,\dots)],[(x_1,\dots,x_{n-1},x_n-1,x_{n+1},\dots)]} = rx_n,$$

$$q_{[(x_1,x_2,\dots)],[(1,x_1,\dots,x_{n-1},x_n-1,x_{n+1},\dots)]} = mx_n,$$

$$q_{[(x_1,x_2,\dots)],[(x_1,x_2,\dots)]} = -(d+r+m)x_n$$

if $x_n > 0$. The Markov chain $(\xi_t)_{t \geq 0}$ has one absorbing state $\mathbf{0} = [(0, 0, 0, \dots)]$ and this state is accessible from the other states. The other states communicate with each other but they are transient (a state is called *transient* if the probability of returning to it is less than one).

The main issue discussed in [97] is the size distribution of paralog families. Let $s_n(t)$ be the number of n-element families at time t, i.e.

$$s_n(t) = \#\{i : x_i(t) = n\},$$

where # denotes the number of elements of the set and let $\mathbf{s}(t) = (s_1(t), s_2(t), \ldots)$. The fundamental evolutionary events change the values of the string $\mathbf{s}(t)$. In this way we obtain another Markov chain but now on the space X. This chain has an intensity matrix Q with the following non-zero entries off the main diagonal:

$$q_{(s_1,s_2,\ldots),(s_1,\ldots,s_{n-1},s_n-1,s_{n+1}+1,s_{n+2},\ldots)} = ns_n d \quad \text{for } n \geq 1,$$

$$q_{(s_1,s_2,\ldots),(s_1,\ldots,s_{n-2},s_{n-1}+1,s_n-1,s_{n+1},\ldots)} = ns_n r \quad \text{for } n \geq 2,$$

$$q_{(s_1,s_2,\ldots),(s_1+1,\ldots,s_{n-2},s_{n-1}+1,s_n-1,s_{n+1},\ldots)} = ns_n m \quad \text{for } n \geq 3,$$

$$q_{(s_1,s_2,\ldots),(s_1-1,s_2,\ldots)} = s_1 r,$$

$$q_{(s_1,s_2,\ldots),(s_1+2,s_2-1,s_3,\ldots)} = 2s_2 m.$$

We will explain the last of these formulas—the reader can easily check the others by himself. Since we have s_2 two-element families in the second group, the probability of the mutation of one gene from this group at time Δt is

$$2 \times s_2 \times m \times \Delta t + o(\Delta t).$$

When a single gene is mutated, both the mutated gene and the other gene from the same family will now appear in single-element families, hence $s_1 + 2$ appears in the first term. We have obtained an interesting Markov chain on the space X. As in the previous example, this chain has one absorbing state $\mathbf{0} = (0, 0, \ldots)$ and this state is accessible from the other states. The other states communicate with each other but they are transient.

Since both above models are quite sophisticated and difficult to analyse directly, from an application point of view it is important to investigate how the expectation value of $s_n(t)$ behaves, which can be treated as the observed number of n-element families in the genome. Let $\bar{s}_n(t) = \mathrm{E}\, s_n(t)$. It can be easily deduced that

$$\bar{s}_1' = -(d+r)\bar{s}_1 + 2(2m+r)\bar{s}_2 + m \sum_{j=3}^{\infty} j\bar{s}_j, \tag{1.16}$$

$$\bar{s}_n' = d(n-1)\bar{s}_{n-1} - (d+r+m)n\bar{s}_n + (r+m)(n+1)\bar{s}_{n+1}, \quad n \geq 2.$$

The following theorem describes the asymptotic behaviour of the sequences $\bar{s}(t) = (\bar{s}_1(t), \bar{s}_2(t), \ldots)$ as $t \to \infty$.

1.3 Examples

Theorem 1.2 ([97]) *Let l_w^1 be the space of sequences $(\bar{s}_n)$ satisfying condition $\sum_{n=1}^{\infty} n|\bar{s}_n| < \infty$. Then there exists a sequence $(\bar{s}_n^*) \in l_w^1$ such that for every solution $(\bar{s}_n(t))$ of (1.16) with initial condition $(\bar{s}_n(0)) \in l_w^1$ we have*

$$\lim_{t \to \infty} e^{(r-d)t} \bar{s}_n(t) = C \bar{s}_n^* \tag{1.17}$$

for each $n = 1, 2, \ldots$ and for some C dependent on $(\bar{s}_n(0))$. Moreover, if $d = r$, then

$$\lim_{t \to \infty} \bar{s}_n(t) = C n^{-1} \left(\frac{r}{r+m} \right)^n. \tag{1.18}$$

Chapter 2
Limit Passages

In this chapter we introduce the mathematical apparatus needed to maintain macroscopic models from individual models through appropriate limit passages. We restrict our investigation to pure-jump IBMs. We show that starting from such models we can obtain in limits: phenotype models with intra-species competition, McKenrick's age-structure model, and a superprocess which is a simplified version of Dawson's superprocess.

2.1 From IBMs to Macroscopic Models: Introduction

We will deal with the following issue. Let us consider a sequence of individual models with features from the set $\mathcal{X}$. Let $\mathcal{M}$ be a space of finite Borel measures on the set $\mathcal{X}$ with the topology of weak convergence. We assume that the sequence of measures $(\frac{1}{N}\mu_N)$, $\mu_N \in \mathcal{N}$, is weakly convergent to the measure $\mu \in \mathcal{M}$. Let (ξ_t^N) be a process with values in $\mathcal{N}$, satisfying the initial condition $\xi_0^N = \mu_N$ and describing the evolution of the population state in the N-th model. What can be said about the sequence of stochastic processes $(\frac{1}{N}\xi_t^N)$ when $N \to \infty$? We expect that, under certain assumptions, the sequence of these processes will converge in some sense to a process (ξ_t) with values in a space of measures.

The following questions arise here. What properties does the process (ξ_t) have? When is it a deterministic process? A process (ξ_t) is called *deterministic* if for a given initial value of $\xi_0 = \mu_0$, $\mu_0 \in \mathcal{M}$, there are measures μ_t, $t \geq 0$, depending only on μ_0 such that $\xi_t = \mu_t$. Can the evolution of measures obtained in this way

---

Electronic Supplementary Material The online version of this chapter (https://doi.org/10.1007/978-3-031-75270-4_2) contains supplementary material, which is available to authorized users.

© The Author(s), under exclusive license to Springer Nature Switzerland AG 2024
R. Rudnicki, R. Wieczorek, *Individual-Based Models and Their Limits*,
SpringerBriefs in Mathematical Methods,
https://doi.org/10.1007/978-3-031-75270-4_2

be described by some transport equation? If the limit process is not deterministic, how can it be described? It is worth emphasising that in this case the state space of the process is a measure space, unlike the situations we have considered so far, where the processes had values in $\mathbb{R}^d$. Why are the answers to these questions so important? Many structured models, i.e. those describing distributions of some parameters in a population are derived *ad hoc*, which can lead to errors. The individual description is based on the most basic mechanisms occurring in the population, so, on the one hand, we should expect a correct description of the structured model through a limit passage, while at the same time we see what kind of scaling has been applied in this passage. This way of proceeding is similar to that in physics, when using the description of the particle collision process we derive the Boltzmann equation of velocity distribution and the heat equation.

The simplest structures have individual models defined by pure-jump Markov processes, in particular by continuous-time Markov chains. Examples of such models were presented in Sect. 1.3. In Sect. 2.3, we study the limit passages for relatively simple jump models.

In the next chapter we will learn about individual models whose limits are *superprocesses*, i.e. stochastic processes with value in the space of measures. The first of these is the Dawson–Watanabe superprocess. It arises as a limit for a branching Brownian motion—a combination of the movement of individuals and a birth-death process. We also present the historical superprocess and the Fleming–Viot superprocess related to the evolution of the distribution of alleles in a population.

In Chap. 5 we study the model evolution of the phenotype distribution. In this case, the only changes in the population structure occur at random times of birth and death of individuals. In Chap. 6 we present more complex models for the evolution of phytoplankton distributions. In such models, phytoplankton cells or aggregates move according to Brownian motion, so the output model will be described by both stochastic equations and jump processes such as birth and death or coagulation, fragmentation and growth in the case of aggregation models.

In order to give an idea of the issues we present a relatively simple example of a limit passage in an individual model in which there is no interaction between individuals and no birth and death processes.

Example 2.1 Consider a population consisting of N individuals moving in the space $\mathbb{R}^d$ according to the following system of stochastic equations

$$dX_t^i = b(X_t^i)\,dt + \sigma(X_t^i)\,dW_t^i, \tag{2.1}$$

where X_t^i is the location of the i-th individual. We assume that the initial position X_0^i of all individuals is the same. We assume that the number of individuals in the population has not changed during the time interval and that the individuals move independently of each other. The distribution of the position of the entire population

2.2 From a Jump Process to Diffusion

at time t is represented by the following Markov process with values in the space of probabilistic measures $\mathcal{M}_1 = \mathcal{M}_1(\mathbb{R}^d)$:

$$\xi_t^N = \frac{1}{N}\left(\delta_{X_t^1} + \cdots + \delta_{X_t^N}\right). \tag{2.2}$$

It is worth noting that for a fixed t the expression ξ_t^N is a random variable with values in the measure space $\mathcal{M}_1$. A random variable with values in the measure space is called *random measure*. We are interested in the limit $\lim_{N\to\infty}\xi_t^N$. Note that for any fixed set $B \in \mathcal{B}(\mathbb{R}^d)$ and $t \geq 0$

$$\xi_t^N(B) = \frac{1}{N}\#\{i : X_t^i \in B,\ i \leq N\} = \frac{1}{N}\sum_{i=1}^{N}\mathbf{1}_B(X_t^i),$$

hence ξ_t^N is the empirical measure for the sequence of random variables $X_t^1,\ldots,X_t^N$. Since the random variables $X_t^1,\ldots,X_t^N$ are independent and identically distributed, from the law of large numbers it follows that $\xi_t^N(B) \to P(X_t^1 \in B)$ a.e. Let $\xi_t(B) = P(X_t^1 \in B)$. Then ξ_t is a measure and the map $t \mapsto \xi_t$ is a deterministic process. Thus, we have shown that at any fixed t the measures of ξ_t^N are *pointwise convergent* a.e. to the measure ξ_t. We used the term pointwise convergence here because measures are functions defined on sets, so the points for measures are measurable sets.

In the above example, we considered a rather specific convergence of processes $(\xi_t^N)_{t\geq 0}$ to a deterministic process $(\xi_t)_{t\geq 0}$. In further considerations we will be interested in the weak convergence of measures (convergence of their distributions) and the convergence of the distributions of processes $(\xi_t^N)_{t\geq 0}$ to the distribution of the limit process $(\xi_t^N)_{t\geq 0}$. It is worth noting in which spaces weak convergence is studied. In the example considered, the processes $(\xi_t^N)_{t\geq 0}$ took values in the space of finite Borel measures $\mathcal{M} = \mathcal{M}(\mathbb{R}^d)$. Thus, at a fixed t, the distribution of the random variable ξ_t^N is a measure from the space $\mathcal{M}_1(\mathcal{M}, \mathcal{B}(\mathcal{M}))$. The process $(\xi_t^N)_{t\geq 0}$ can be treated as a random variable with values in the Skorohod space $\mathbb{D} = \mathbb{D}([0, \infty), \mathcal{M})$, and its distribution is a measure from the space $\mathcal{M}_1(\mathbb{D}, \mathcal{B}(\mathbb{D}))$. All spaces listed here are metrizable and, with appropriately chosen metrics, are Polish spaces (see Appendix B).

2.2 From a Jump Process to Diffusion

We now give a limit passage theorem from a jump process to a diffusion process which we will use later.

We consider a jump process on $\mathbb{R}^d$. We define such a process by giving the jump intensity from a point $\mathbf{x}$ to the set B. For a fixed point $\mathbf{x} \in \mathbb{R}^d$, the jump intensity is the measure $K(\mathbf{x}, B)$ such that the probability of moving from a point $\mathbf{x}$ to a set B in the time interval $[t, t + \Delta t]$ is $K(\mathbf{x}, B)\Delta t + o(\Delta t)$. The jump process has the infinitesimal generator of the form

$$A f(\mathbf{x}) = \int (f(\mathbf{y}) - f(\mathbf{x})) K(\mathbf{x}, d\mathbf{y}).$$

We assume that $K(\mathbf{x}, \{\mathbf{x}\}) = 0$ for $\mathbf{x} \in \mathbb{R}^d$ and $\sup_{\mathbf{x} \in \mathbb{R}^d} K(\mathbf{x}, \mathbb{R}^d) < \infty$. The second assumption guarantees a finite number of jumps on a bounded time interval, and therefore the existence of a process. This assumption can be weakened by assuming that the process is non-explosive.

We consider a diffusion process with a generator

$$Lg(\mathbf{x}) = \sum_{i=1}^n b^i(\mathbf{x}) \frac{\partial g(\mathbf{x})}{\partial x^i} + \frac{1}{2} \sum_{i,j=1}^n a^{ij}(\mathbf{x}) \frac{\partial^2 g(\mathbf{x})}{\partial x^i \partial x^j}, \qquad (2.3)$$

where b^i and a^{ij} are continuous functions. We assume that for every initial condition X_0 the martingale problem (see Sec. A.5)

$$f(X_t) - f(X_0) - \int_0^t Lf(X_r)\,dr \quad \text{for } f \in \mathfrak{D}(L) \qquad (2.4)$$

has a unique solution. We consider a sequence of jump processes $(X_t^N)_{t \geq 0}$ with values in $\mathbb{R}^d$ and kernels of the jump operator $K^N(\mathbf{x}, d\mathbf{y})$. We assume that the map:

$$\mathbf{x} \mapsto \int_{\mathbb{R}^d} f(\mathbf{y}) K^N(\mathbf{x}, d\mathbf{y}) \qquad (2.5)$$

is continuous for continuous functions $f : \mathbb{R}^d \to [0, \infty)$ converging to one in infinity, vanishing at zero and

$$\int_{\mathbb{R}^d} \|\mathbf{y}\|^2 K^N(\mathbf{x}, d\mathbf{y}) < \infty. \qquad (2.6)$$

Let

$$b^N(\mathbf{x}) = \int_{\mathbb{R}^d} (\mathbf{y} - \mathbf{x}) K^N(\mathbf{x}, d\mathbf{y}),$$

$$a^N(\mathbf{x}) = \int_{\mathbb{R}^d} (\mathbf{y} - \mathbf{x})^T (\mathbf{y} - \mathbf{x}) K^N(\mathbf{x}, d\mathbf{y}).$$

We assume that

$$b^N(\mathbf{x}) \to b(\mathbf{x}), \quad a^N(\mathbf{x}) \to a(\mathbf{x}) \quad \text{almost uniformly.} \tag{2.7}$$

Moreover, we assume that

$$\forall_{r>0} \forall_{\varepsilon>0} \lim_{N\to\infty} \sup_{\{\mathbf{x}:\, \|\mathbf{x}\|\leq r\}} \int \|\mathbf{y} - \mathbf{x}\|^2 \mathbf{1}_{\{\|\mathbf{y}-\mathbf{x}\|>\varepsilon\}}(\mathbf{y}) K^N(\mathbf{x}, d\mathbf{y}) = 0. \tag{2.8}$$

Theorem 2.1 *Let $(X_t^N)_{t\geq 0}$ be a sequence of jump processes with kernels K^N satisfying conditions (2.5)–(2.8). If the sequence X_0^N is weakly convergent to X_0, then the distributions of the processes (X_t^N) are weakly convergent to the distribution of the diffusion process (X_t) satisfying the martingale problem (2.4).*

The proof of Theorem 2.1 is given in [59] Ch. 9, Th. 4.21. Theorem 2.1 follows also from Theorem B.8 on the convergence of Feller processes.

Example 2.2 Consider a sequence of walks on $\mathbb{R}$ such that we jump in the N-th process from point x to points $x - \frac{1}{N}$ and $x + \frac{1}{N}$ with intensities $\alpha |x| N^2$. Then

$$K^N(x, \{x - 1/N\}) = K^N(x, \{x + 1/N\}) = \alpha |x| N^2$$

and $K^N(x, B) = 0$ if $(x - 1/N) \notin B$ and $(x + 1/N) \notin B$. Then

$$b^N(x) = \left(\frac{1}{N} - \frac{1}{N}\right)\alpha|x|N^2 = 0, \quad a^N(x) = \left(\frac{1}{N^2} + \frac{1}{N^2}\right)\alpha|x|N^2 = 2\alpha|x|.$$

Condition (2.8) is satisfied because the integral expression is equal to zero when $1/N < \varepsilon$. Thus the sequence of processes is convergent in distribution to the diffusion process with generator $Lg(x) = \alpha|x|g''(x)$, i.e. the limit process $(\xi_t)_{t\geq 0}$ satisfies the following stochastic equation

$$d\xi_t = \sqrt{2\alpha|\xi_t|}\, dW_t. \tag{2.9}$$

2.3 Examples of Limit Passages

We now give some examples of the limits of sequences of pure jump individual models. We assume that the N-th model is a Markov process (ξ_t^N) with values in the measure space $\mathcal{N}_N = \{\mu = \frac{1}{N}\nu : \nu \in \mathcal{N}\}$ equipped with a σ-algebra of Borel sets $\mathcal{B}(\mathcal{N}_N)$ in the topology of weak convergence. The process (ξ_t^N) is defined by giving intensity $K_N(\mu, B)$ of a jump from a measure μ to a set $B \in \mathcal{B}(\mathcal{N}_N)$. We recall that the function $B \mapsto K_N(\mu, B)$ is a measure for each measure μ, the function $\mu \mapsto K_N(\mu, B)$ is measurable for each set $B \in \mathcal{B}(\mathcal{N}_N)$, and $K_N(\mu, \{\mu\}) = 0$. The infinitesimal generator of the jump process (ξ_t^N) is of the form

$$L_N F(\mu) = \int_{\mathcal{N}_N} (F(\nu) - F(\mu)) K_N(\mu, d\nu),$$

where $F: \mathcal{N}_N \to \mathbb{R}$ is a continuous and bounded function.

The question of existence of a process (ξ_t^N) for a given jump intensity function K is studied similarly to the case of Markov chains with continuous time on a countable state space. It comes down to checking whether the number of jumps of the process on a finite time interval $[0, T]$ is finite a.e. For example, if the function K is bounded, then it is, but in some applications there appear unbounded functions K and then the proof that the process is non-explosive is more difficult. We consider processes that are well defined.

We are interested in the question of convergence in distribution of a sequence of processes (ξ_t^N) to some limit process (ξ_t). Before going on to discuss this issue, it is worth noting that the study of other individual models can often be reduced to the study of jump models. For example, when individuals move in the space $\mathbb{R}^d$ according to the equation $\mathbf{x}'(t) = \mathbf{b}(\mathbf{x})$, their motion can be approximated by a jump process. If $\mu = \frac{1}{N} \sum \delta_{\mathbf{x}_i}$, then in the N-th process we add jumps with intensity N from a measure μ to the measures ν of the form

$$\nu = \mu + \frac{1}{N}\delta_{\mathbf{x}_i + \mathbf{b}(\mathbf{x}_i)/N} - \frac{1}{N}\delta_{\mathbf{x}_i}. \tag{2.10}$$

We can do the same for diffusion, for example, by replacing d-dimensional Brownian motion process with a pure-jump process in which in the N-th process we jump with intensity $N/2$ from a measure μ to the measure ν defined by the formula

$$\nu = \mu + \frac{1}{N}\delta_{\mathbf{x}_i + \boldsymbol{\varepsilon}_k/\sqrt{N}} - \frac{1}{N}\delta_{\mathbf{x}_i}, \tag{2.11}$$

where a sequence $\boldsymbol{\varepsilon}_1, \ldots, \boldsymbol{\varepsilon}_{2d}$ contains all vectors of the form $\pm[0, \ldots, 0, 1, 0, \ldots, 0]$.

We will consider individual-based models, in which the set of traits $\mathcal{X}$ is some regular subset of $\mathbb{R}^d$, for example $\mathcal{X}$ is a domain or a closed set with a non-empty interior and a smooth boundary. For simplicity of notation, we write $\langle f, \mu \rangle$ instead of $\int f(x)\, \mu(dx)$. The symbol C_c^∞ denotes the space of C^∞-functions $h: \mathcal{X} \to \mathbb{R}$ with compact supports. For any measurable bounded function h we write $F_h(\mu) = \exp\langle h, \mu \rangle$ and $G_h(\mu) = \langle h, \mu \rangle$ for $\mu \in \mathcal{M}$.

The study of the convergence of a sequence (ξ_t^N) to some process (ξ_t) with values in the space $\mathcal{M}$ is carried out in three steps:

1. Checking that the sequence of processes (ξ_t^N) is tight, which in suitable spaces (see Section B3) implies the weak compactness of the sequence of distributions of processes (ξ_t^N).

2.3 Examples of Limit Passages

2. Determination of the limiting martingale problem.
3. Verification that the martingale problem has a unique solution.

Although the general methods for checking convergence are already quite well developed, the investigation of specific individual models is still a technically quite advanced issue. The study of the tightness of a sequence of distributions of processes (ξ_t^N) with values in the space S is usually done using Jakubowski's criterion [61] of tightness in spaces $\mathbb{D}([0, \infty), S)$ and showing that some moment of the distributions is bounded (see also [10] and Section B3). In the second point we determine the generator of the limit process and we will devote the rest of the discussion to this issue.

Let $(\xi_t)_{t \geq 0}$ be a homogeneous Markov process with values in the space of finite measures $\mathcal{M}$ (or its subspace, e.g. $\mathcal{N}_N$) and let $\mathcal{P} : [0, \infty) \times \mathcal{M} \times \mathcal{B}(\mathcal{M}) \to [0, 1]$ be the transition probability function for this process. The process $(\xi_t)_{t \geq 0}$ is associated with a semigroup of operators $\{U(t)\}_{t \geq 0}$ on the space of measurable and bounded functions $B(\mathcal{M}, \mathcal{B}(\mathcal{M}))$ defined by the formula

$$U(t) F(\mu) = \mathbb{E}_\mu F(\xi_t) = \int_{\mathcal{M}} F(\nu) \, \mathcal{P}(t, \mu, d\nu)$$

and an infinitesimal generator L defined on the set $\mathfrak{D}(L) \subset B(\mathcal{M}, \mathcal{B}(\mathcal{M}))$ containing all functions F such that there exists a limit

$$LF(\mu) = \lim_{t \to 0^+} \frac{U(t) F(\mu) - F(\mu)}{t} = \lim_{t \to 0^+} \frac{1}{t} \left(\int_{\mathcal{M}} (F(\nu) - F(\mu)) \, \mathcal{P}(t, \mu, d\nu) \right),$$

and the convergence is uniform on $\mathcal{M}$ (see Sec. B.5). In particular

$$LG_h(\mu) = \lim_{t \to 0^+} \frac{1}{t} \left(\int_{\mathcal{M}} \langle h, \nu - \mu \rangle \, \mathcal{P}(t, \mu, d\nu) \right), \tag{2.12}$$

$$LF_h(\mu) = e^{\langle h, \mu \rangle} \lim_{t \to 0^+} \frac{1}{t} \left(\int_{\mathcal{M}} \left(e^{\langle h, \nu - \mu \rangle} - 1 \right) \mathcal{P}(t, \mu, d\nu) \right). \tag{2.13}$$

We now suppose that the process $(\xi_t)_{t \geq 0}$ is deterministic, and even more, that there exists a semiflow $(\Pi_t)_{t \geq 0}$ defined on the space of measures $\mathcal{M}$ such that $\xi_t = \Pi_t \xi_0$. Then

$$LG_h(\mu) = \lim_{t \to 0^+} \frac{1}{t} \langle h, \Pi_t \mu - \mu \rangle, \tag{2.14}$$

$$LF_h(\mu) = e^{\langle h, \mu \rangle} \lim_{t \to 0^+} \frac{1}{t} \left(e^{\langle h, \Pi_t \mu - \mu \rangle} - 1 \right). \tag{2.15}$$

In determining the limits of individual models we usually obtain explicit formulae for the expressions $LG_h(\mu)$ and $LF_h(\mu)$. Using the formulae (2.12)–(2.15) we can often find the evolution of the distribution of the limit process. Note that the set $\{G_h\colon h \in C_c^\infty\}$ is too small in the space $C_b(\mathscr{M})$ to uniquely reconstruct the distribution of the process $(\xi_t)_{t\geq 0}$ (see Example 2.8, in which the processes $(\xi_t^N)_{t\geq 0}$ are not constant, and $L_N G_h \equiv 0$ for any h). The set $\{F_h\colon h \in C_c^\infty\}$ is linearly dense in the space $C_b(\mathscr{M})$ (it follows from a version of the Stone–Weierstrass theorem, see Section B.5 for a detailed discussion). Therefore, the distribution of the limit process, if such a process exists, is determined uniquely by giving the value of the function LF_h for $h \in C_c^\infty$. If for example

$$LF_h(\mu) = \langle Ah, \mu \rangle \exp\langle h, \mu \rangle, \qquad (2.16)$$

where A is a linear operator on the space C_c^∞ and there is a semiflow $(\Pi_t)_{t\geq 0}$ on the space $\mathscr{M}$ such that for every $h \in C_c^\infty$

$$\langle Ah, \mu \rangle = \lim_{t\to 0^+} \frac{1}{t} \langle h, \Pi_t \mu - \mu \rangle \qquad (2.17)$$

and the limit is uniform on $\mathscr{M}$, then $\xi_t = \Pi_t \xi_0$. The semiflow $(\Pi_t)_{t\geq 0}$ is linear, i.e.

$$\Pi_t(c_1\mu_1 + c_2\mu_2) = c_1\Pi_t\mu_1 + c_2\Pi_t\mu_2$$

for $\mu_1, \mu_2 \in \mathscr{M}$ and $c_1, c_2 > 0$. The trajectories of the semiflow $(\Pi_t)_{t\geq 0}$ are solution of the following integral equation

$$\langle h, \Pi_t \mu \rangle = \langle h, \mu \rangle + \int_0^t \langle Ah, \Pi_s \mu \rangle\, ds \quad \text{for all } h \in C_c^\infty. \qquad (2.18)$$

A semiflow $(\Pi_t)_{t\geq 0}$ on the space of measures $\mathscr{M}$ can also be the limit process when the operator A is nonlinear or when A depends on $\mathscr{M}$ (see Example 2.6). In this case it is convenient to use the Eq. (2.18). The resulting semiflow will no longer be linear.

Example 2.3 Let $\mathbf{b}\colon \mathbb{R}^d \to \mathbb{R}^d$ be a C^1-function such that $\|\mathbf{b}(\mathbf{x})\| \leq \alpha + \beta\|\mathbf{x}\|$ for $\mathbf{x} \in \mathbb{R}^d$ for some positive constants α, β. We consider a sequence of processes (ξ_t^N) such that in the N-th process we jump with intensity N from the measure $\mu = \frac{1}{N} \sum \delta_{\mathbf{x}_i}$ to any measure ν defined by (2.10). Then

$$K_N(\mu, \{\mu + \frac{1}{N}\delta_{\mathbf{x}_i + \mathbf{b}(\mathbf{x}_i)/N} - \frac{1}{N}\delta_{\mathbf{x}_i}\}) = N$$

for all i and $K_N(\mu, B) = 0$, if none of the measures $\mu + \frac{1}{N}\delta_{\mathbf{x}_i + \mathbf{b}(\mathbf{x}_i)/N} - \frac{1}{N}\delta_{\mathbf{x}_i}$ does not belong to the set B. The infinitesimal generator of the process (ξ_t^N) is of the form

$$L_N F(\mu) = \sum_i N\left(F\left(\mu + \frac{1}{N}\delta_{\mathbf{x}_i + \mathbf{b}(\mathbf{x}_i)/N} - \frac{1}{N}\delta_{\mathbf{x}_i}\right) - F(\mu)\right).$$

Observe that

2.3 Examples of Limit Passages

$$L_N F_h(\mu) = \sum_i N\left(\exp\langle h, \mu + \frac{1}{N}\delta_{\mathbf{x}_i + \mathbf{b}(\mathbf{x}_i)/N} - \frac{1}{N}\delta_{\mathbf{x}_i}\rangle - \exp\langle h, \mu\rangle\right)$$

$$= \sum_i N\left(\exp\left(\frac{1}{N}\bigl(h(\mathbf{x}_i + \mathbf{b}(\mathbf{x}_i)/N) - h(\mathbf{x}_i)\bigr)\right) - 1\right)\exp\langle h, \mu\rangle$$

$$= \sum_i N\left(\exp\left(\frac{1}{N^2}\operatorname{grad} h(\mathbf{x}_i)\cdot \mathbf{b}(\mathbf{x}_i) + o(N^{-2})\right) - 1\right)\exp\langle h, \mu\rangle$$

$$= \sum_i \left(\frac{1}{N}\operatorname{grad} h(\mathbf{x}_i)\cdot \mathbf{b}(\mathbf{x}_i) + o(N^{-1})\right)\exp\langle h, \mu\rangle$$

$$= \langle \mathbf{b}\cdot \operatorname{grad} h, \mu\rangle \exp\langle h, \mu\rangle + o(1).$$

We have checked that the sequence $L_N F_h(\mu)$ has the limit $\langle Ah, \mu\rangle \exp\langle h, \mu\rangle$, where $Ah = \mathbf{b}\cdot \operatorname{grad} h$. The solutions of the differential equation $\mathbf{x}' = \mathbf{b}(\mathbf{x})$ define a flow $(\pi_t)_{t\in\mathbb{R}}$ on the space $\mathbb{R}^d$, and this in turn determines the flow $(\Pi_t)_{t\in\mathbb{R}}$ on the space of measures $\mathscr{M}$ defined by $\Pi_t\mu(B) = \mu(\pi_{-t}(B))$ for $\mu \in \mathscr{M}$, $B \in \mathscr{B}(\mathbb{R}^d)$. We check the formula (2.17). Let $\{P^*(t)\}_{t\in\mathbb{R}}$ be a group of Koopman operators for the flow $(\pi_t)_{t\in\mathbb{R}}$, i.e. $P^*(t)f(\mathbf{x}) = f(\pi_t\mathbf{x})$ for $f \in C_0(\mathbb{R}^d)$. Then

$$\langle h, \Pi_t\mu\rangle = \int_{\mathbb{R}^d} h(\mathbf{x})\,\Pi_t\mu(d\mathbf{x}) = \int_{\mathbb{R}^d} h(\pi_t(\mathbf{x}))\,\mu(d\mathbf{x}) = \langle P^*(t)h, \mu\rangle.$$

Since the operator $Ah = \mathbf{b}\cdot \operatorname{grad} h$ is the generator of the group $\{P^*(t)\}_{t\in\mathbb{R}}$, we have

$$\lim_{t\to\infty} \frac{1}{t}\langle h, \Pi_t\mu - \mu\rangle = \langle Ah, \mu\rangle.$$

Thus the limit process is defined by the formula $\xi_t = \Pi_t\xi_0$.

Example 2.4 We consider a sequence of processes (ξ_t^N) such that in the N-th process we jump with intensity $\frac{1}{2}N$ from the measure μ to any measure ν defined by (2.11). In this case

$$L_N F_h(\mu) = \sum_i \sum_{k=1}^{2d} \tfrac{1}{2} N\left(\exp\langle h, \mu + \frac{1}{N}\delta_{\mathbf{x}_i + \mathbf{e}_k/\sqrt{N}} - \frac{1}{N}\delta_{\mathbf{x}_i}\rangle - \exp\langle h, \mu\rangle\right)$$

$$= \sum_i \sum_{k=1}^{2d} \tfrac{1}{2} N\left(\exp\left(\frac{1}{N}\bigl(h(\mathbf{x}_i + \mathbf{e}_k/\sqrt{N}) - h(\mathbf{x}_i)\bigr)\right) - 1\right)\exp\langle h, \mu\rangle.$$

Observe that if $\varepsilon_k = [0, \ldots, 0, 1, 0, \ldots, 0]$, where the 1 is on the k-th position, then

$$h(\mathbf{x}_i + \varepsilon_k/\sqrt{N}) - h(\mathbf{x}_i) = \frac{1}{\sqrt{N}} \frac{\partial h}{\partial x_k}(\mathbf{x}_i) + \frac{1}{2N} \frac{\partial^2 h}{\partial x_k^2}(\mathbf{x}_i) + o(N^{-1}),$$

and consequently

$$\exp\frac{1}{N}\left(h(\mathbf{x}_i + \varepsilon_k/\sqrt{N}) - h(\mathbf{x}_i)\right) - 1 = \frac{1}{N\sqrt{N}} \frac{\partial h}{\partial x_k}(\mathbf{x}_i) + \frac{1}{2N^2} \frac{\partial^2 h}{\partial x_k^2}(\mathbf{x}_i) + o(N^{-2}).$$

Analogously, if $\varepsilon_{d+k} = -\varepsilon_k$, then

$$\exp\frac{1}{N}\left(h(\mathbf{x}_i + \varepsilon_{d+k}/N) - h(\mathbf{x}_i)\right) - 1 = -\frac{1}{N\sqrt{N}} \frac{\partial h}{\partial x_k}(\mathbf{x}_i) + \frac{1}{2N^2} \frac{\partial^2 h}{\partial x_k^2}(\mathbf{x}_i) + o(N^{-2}).$$

By combining in the formula for $L_N F_h(\mu)$ the expressions containing ε_k and ε_{d+k} we obtain

$$L_N F_h(\mu) = \sum_i \sum_{k=1}^d \left(\frac{1}{2N} \frac{\partial^2 h}{\partial x_k^2}(\mathbf{x}_i) \exp\langle h, \mu\rangle + o(N^{-1})\right)$$

$$= \sum_i \left(\frac{1}{2N} \Delta h(\mathbf{x}_i) \exp\langle h, \mu\rangle + o(N^{-1})\right)$$

$$= \langle \tfrac{1}{2}\Delta h, \mu\rangle \exp\langle h, \mu\rangle + o(1).$$

As the limit we obtain the mapping LF_h of the form (2.16) with $Ah = \tfrac{1}{2}\Delta h$. Also in this example, the limit process is of the form $\xi_t = \Pi_t \xi_0$, and the semiflow $(\Pi_t)_{t \geq 0}$ is defined by $\Pi_t \mu(dx) = u(t, x)\,dx$, where $u(t, x)$ is a solution of the heat equation $\frac{\partial u}{\partial t} = \tfrac{1}{2}\Delta u(t, x)$, such that μ is a weak limit of measures $u(t, x)\,dx$ as $t \to 0^+$.

Example 2.5 Consider a model describing the birth and death process in which birth rates $b(x)$ and death rates $d(x)$ depend on the phenotype $x \in \mathcal{X}$. We assume that the set of traits $\mathcal{X}$ is a Polish space, for example $\mathcal{X}$ is a closed subset of $\mathbb{R}^d$, with the Σ-algebra $\Sigma = \mathcal{B}(\mathcal{X})$ of Borel subsets. We assume that the functions b and d are continuous, non-negative and bounded. The N-th process (ξ_t^N) jumps from the measure $\mu = \frac{1}{N}\sum_i \delta_{x_i}$ to the measure $\mu + \frac{1}{N}\delta_{x_i}$ with intensity $b(x_i)$ and to the measure $\mu - \frac{1}{N}\delta_{x_i}$ with intensity $d(x_i)$. Then for any continuous and bounded function $F: \mathcal{N}_N \to \mathbb{R}$ we have

$$L_N F(\mu) = \sum_i \left(b(x_i)\bigl(F(\mu + \tfrac{1}{N}\delta_{x_i}) - F(\mu)\bigr) + d(x_i)\bigl(F(\mu - \tfrac{1}{N}\delta_{x_i}) - F(\mu)\bigr)\right).$$

2.3 Examples of Limit Passages

Hence, for any measurable and bounded function $h\colon \mathscr{X} \to \mathbb{R}$ we have

$$L_N F_h(\mu) = \sum_i \left(b(x_i)e^{h(x_i)/N} - b(x_i) + d(x_i)e^{-h(x_i)/N} - d(x_i)\right) \exp\langle h, \mu\rangle$$

$$= \sum_i \left(\tfrac{1}{N}b(x_i)h(x_i) - \tfrac{1}{N}d(x_i)h(x_i) + o(N^{-1})\right) \exp\langle h, \mu\rangle$$

$$= \left(\langle (b-d)h, \mu\rangle + o(1)\right) \exp\langle h, \mu\rangle,$$

thus $L_N F_h(\mu) \to \langle (b-d)h, \mu\rangle \exp\langle h, \mu\rangle$, as $N \to \infty$. Also in this example we obtain in the limit a mapping LF_h of the form (2.16) with $Ah(x) = (b(x) - d(x))h(x)$. The limit process is of the form $\xi_t = \Pi_t \xi_0$ and the flow $(\Pi_t)_{t\geq 0}$ is defined by $\Pi_t \mu(dx) = e^{(b(x)-d(x))t} \mu(dx)$. It is worth noting that the trajectories of the flow $(\Pi_t)_{t\geq 0}$ are continuous function also in the total variation norm: $\|\mu\|_{TV} = |\mu|(\mathscr{X})$.

Example 2.6 Let us consider a slightly more complicated version of the model of Example 2.5, in which the mortality rate still depends on intra-species competition. We assume that in the N-th model the death rate of the i-th individual is

$$\widetilde{d}(x_i) = d(x_i) + \frac{1}{N}\sum_j U(x_i, x_j). \tag{2.19}$$

The function U is non-negative, continuous and bounded. Then, in calculating $L_N F_h(\mu)$, the expression $d(x_i)$ is replaced by $\widetilde{d}(x_i)$. In this way we obtain

$$L_N F_h(\mu) = \sum_i \left(\tfrac{1}{N}b(x_i) - \tfrac{1}{N}d(x_i) - \tfrac{1}{N^2}\sum_j U(x_i, x_j)\right)h(x_i) \exp\langle h, \mu\rangle + o(1).$$

Since

$$\frac{1}{N^2} \sum_i \sum_j U(x_i, x_j) h(x_i) = \int_{\mathscr{X}} \int_{\mathscr{X}} U(x, y) h(x)\, \mu(dx)\, \mu(dy),$$

so assuming that $\mathscr{U}(x, \mu) = \int_{\mathscr{X}} U(x, y)\, \mu(dy)$ we obtain

$$LF_h(\mu) = \langle (b - d - \mathscr{U}(\cdot, \mu))h, \mu\rangle \exp\langle h, \mu\rangle.$$

The limit of the processes (ξ_t^N) is also a semiflow $(\Pi_t)_{t\geq 0}$ on the space $\mathscr{M}$ which satisfies the following nonlinear integral equation

$$\langle h, \Pi_t \mu\rangle = \langle h, \mu\rangle + \int_0^t \langle (b - d - \mathscr{U}(\cdot, \mu))h, \Pi_s \mu\rangle\, ds \tag{2.20}$$

for all $h \in C_c^\infty$. Let $\mu_t = \Pi_t \mu$. It is easy to check that the function $t \mapsto \mu_t$ satisfies the following ordinary differential equation

$$\frac{d\mu_t}{dt} = \Big(b(x) - d(x) - \int_{\mathcal{X}} U(x, y)\, \mu_t(dy)\Big) \mu_t \tag{2.21}$$

on the space $\mathcal{M}$ equipped with *total variation metric* $d_{TV}(\mu, \nu) = \|\mu - \nu\|_{TV}$. The set $\mathcal{M}$ is a closed and convex subset in a Banach space of finite signed measures $\mathcal{M}_s$ and therefore we can treat this equation as a nonlinear ordinary differential equation in a Banach space. We obtain local existence and uniqueness of solutions of the equation by writing this equation in an integral form and using Banach's fixed point theorem. Global existence follows from the fact that the nonlinear factor is negative (as in the logistic equation). The property that the limit process $(\xi_t)_{t \geq 0}$ has continuous and even differentiable trajectories, in the total variation metric is quite typical for pure jump models. For example, such a property holds for a rather general phenotype model considered in Chap. 5.

Example 2.7 An interesting individual model describes the birth and death process, in which the age of the individual a is used instead of the phenotype. We assume that the birth rates b and death rates d are non-negative, continuous and bounded functions defined on the interval $[0, \infty)$. Now in the N-th process we jump from the measure $\mu = \frac{1}{N}\sum_i \delta_{a_i}$ to the measure $\mu + \frac{1}{N}\delta_0$ with intensity $\sum_i b(a_i)$ and to the measure $\mu - \frac{1}{N}\delta_{a_i}$ with intensity $d(a_i)$. As the age of the individual increases according to the equation $a' = 1$, we add jumps approximating the growth equation. We therefore assume that with intensity N the process jumps from a measure μ to the measure $\mu + \frac{1}{N}\delta_{a_i+1/N} - \frac{1}{N}\delta_{a_i}$. Then for any continuous and bounded function $F: \mathcal{N}_N \to \mathbb{R}$ we have

$$L_N F(\mu) = \sum_i \Big(b(a_i)\big(F(\mu + \tfrac{1}{N}\delta_0) - F(\mu)\big) + d(a_i)\big(F(\mu - \tfrac{1}{N}\delta_{a_i}) - F(\mu)\big)\Big)$$

$$+ N \sum_i \Big(F(\mu + \tfrac{1}{N}\delta_{a_i+1/N} - \tfrac{1}{N}\delta_{a_i}) - F(\mu)\Big).$$

An easy computations, which we leave to the reader as an exercise, leads to a limit function $LF_h(\mu) = \langle Ah, \mu\rangle \exp\langle h, \mu\rangle$ with the operator A of the form

$$Ah(a) = b(a)h(0) - d(a)h(a) + h'(a). \tag{2.22}$$

The limit process is still a semiflow $(\Pi_t)_{t \geq 0}$ on measures. If the measure μ has a density (not necessarily probabilistic) $u_0(a)$, then the measure $\Pi_t \mu$ has the density $u(t, a)$ which satisfies the system of McKendrick equations

$$\frac{\partial u}{\partial t} + \frac{\partial u}{\partial a} = -d(a)u(t, a),$$

$$u(t, 0) = \int_0^\infty b(a)u(t, a)\, da.$$

2.3 Examples of Limit Passages

The next example shows that the limit of a relatively simple sequence of individual jump models can be a non-trivial process with values in the space of finite measures.

Example 2.8 We change slightly Example 2.5 assuming that in the N-th process we jump from the measure $\mu = \frac{1}{N}\sum_i \delta_{x_i}$ to the measure $\mu - \frac{1}{N}\delta_{x_i}$ with intensity αN and to the measure $\mu + \frac{1}{N}\delta_{x_i}$ also with intensity αN. For any function $F: \mathcal{N}_N \to \mathbb{R}$ we have

$$L_N F(\mu) = \sum_i \alpha N \left(F\left(\mu + \tfrac{1}{N}\delta_{x_i}\right) + F\left(\mu - \tfrac{1}{N}\delta_{x_i}\right) - 2F(\mu) \right).$$

Let $h: \mathcal{X} \to \mathbb{R}$ be a bounded and measurable function. Then

$$L_N G_h(\mu) = \sum_i \alpha N \left(\langle h, \mu + \tfrac{1}{N}\delta_{x_i} \rangle + \langle h, \mu - \tfrac{1}{N}\delta_{x_i} \rangle - 2\langle h, \mu \rangle \right) = 0.$$

Since $L_N G_h(\mu) = 0$ for any $h \in B(\mathcal{X}, \Sigma)$ and $\mu \in \mathcal{N}_N$, the previous examples suggest that the sequence of processes (ξ_t^N) is convergent to a constant process. It turns out that this is not the case. After a standard computations we obtain

$$L_N F_h(\mu) = \sum_i \alpha N \left(e^{h(x_i)/N} + e^{-h(x_i)/N} - 2 \right) \exp\langle h, \mu \rangle$$

$$= \alpha N \sum_i \left(N^{-2} h^2(x_i) + o(N^{-2}) \right) \exp\langle h, \mu \rangle$$

$$= \left(\langle \alpha h^2, \mu \rangle + o(1) \right) \exp\langle h, \mu \rangle,$$

hence $L_N F_h(\mu) \to \langle \alpha h^2, \mu \rangle \exp\langle h, \mu \rangle$, as $N \to \infty$. Thus the generator of the limit process $(\xi_t)_{t \geq 0}$, if it exists, should satisfy the condition

$$L F_h(\mu) = \langle \alpha h^2, \mu \rangle \exp\langle h, \mu \rangle. \tag{2.23}$$

Due to the nonlinear term αh^2 in the formula (2.23), guessing the limit process is not a simple matter. In order to do so, let us fix the measure $\mu = y\delta_x$, $x \in \mathcal{X}$, $y > 0$. Based on the description of the processes (ξ_t^N) we will try to guess the distribution of the process (ξ_t) with initial condition $\xi_0 = \mu$. For simplicity, let us assume that $k = Ny$ is an integer. We can write the measure μ in the form of a sum $\mu = \frac{1}{N}\sum_{i=1}^k \delta_{x_i}$, $x_1 = \cdots = x_k = x$. The intensities of the jump from μ to $\mu + \frac{1}{N}\delta_{x_i}$ and from μ to $\mu - \frac{1}{N}\delta_{x_i}$ are αN, so the sum of the jump intensities from the measure $y\delta_x$ to the measure $(y + \frac{1}{N})\delta_x$ and to the measure $(y - \frac{1}{N})\delta_x$ are $\alpha Nk = \alpha y N^2$. Since only the coefficient at δ_x changes, we can consider only the limit passage for a process with jumps $1/N$ and $-1/N$ with the intensities of each of these jumps $\alpha y N^2$. From Theorem 2.1 we obtain that the limit process (Y_t)

satisfies the equation

$$dY_t = \sqrt{2\alpha Y_t}\,dW_t, \quad Y_0 = y. \tag{2.24}$$

The point 0 is absorbing for this process and

$$P(Y_t = 0) := 1 - P(Y_s > 0 \text{ for } s \in [0,t]) > 0, \quad t > 0.$$

The process $\xi_t = Y_t \delta_x$ has the following property

$$LF_h(\mu) = \lim_{t \to 0^+} \frac{1}{t} \mathrm{E}\left(e^{\langle h, \xi_t \rangle} - e^{\langle h, \mu \rangle}\right) = \lim_{t \to 0^+} \frac{1}{t} \mathrm{E}\left(e^{h(x)Y_t} - e^{h(x)y}\right).$$

The Laplace transform of the process $(Y_t)_{t \geq 0}$ is given by

$$\mathrm{E}\, e^{-\lambda Y_t} = \exp\left(\frac{-\lambda y}{1 + \lambda \alpha t}\right).$$

Setting $\lambda = -h(x)$ we obtain

$$\mathrm{E}\, e^{h(x)Y_t} = \mathrm{E}\, e^{-\lambda Y_t} = \exp\left(\frac{h(x)y}{1 - h(x)\alpha t}\right). \tag{2.25}$$

Hence

$$LF_h(\mu) = \frac{d}{dt} \exp\left(\frac{h(x)y}{1 - h(x)\alpha t}\right)\bigg|_{t=0} = e^{h(x)y} \frac{d}{dt}\left(\frac{h(x)y}{1 - h(x)\alpha t}\right)\bigg|_{t=0}$$

$$= e^{h(x)y} h(x)y\, h(x)\alpha\, (1 - h(x)\alpha t)^{-2}\bigg|_{t=0} = \alpha y h^2(x) e^{\langle h, \mu \rangle}$$

$$= \langle \alpha h^2, \mu \rangle e^{\langle h, \mu \rangle},$$

and the formula (2.23) holds. Let now

$$\mu = y_1 \delta_{x_1} + \cdots + y_n \delta_{x_n}, \tag{2.26}$$

where $x_1, \ldots, x_n$ is a sequence of different elements from the set $\mathscr{X}$ and $y_1, \ldots, y_n > 0$. Since the jumps of the process (ξ_t^N) in the points x_i are independent, the limit process is of the form $\xi_t = Y_t^1 \delta_{x_1} + \cdots + Y_t^n \delta_{x_n}$, where (Y_t^i) fulfills the equation

$$dY_t^i = \sqrt{2\alpha Y_t^i}\,dW_t^i, \quad Y_0^i = y_i, \tag{2.27}$$

and Wiener processes $(W_t^1), \ldots, (W_t^n)$ are independent.

2.3 Examples of Limit Passages

Let us look in more detail at how the process (ξ_t) works if the measure μ is of the form (2.26). To do this, let us define the probability distribution in the space $[0, \infty)^n$ by the formula:

$$P_{\mu,t}(\Gamma) = P\big((Y_t^1, \ldots, Y_t^n) \in \Gamma\big) \quad \text{for } \Gamma \in \mathcal{B}([0, \infty)^n). \tag{2.28}$$

According to the distribution $P_{\mu,t}$ we draw a point $\mathbf{r} = (r_1, \ldots, r_n)$ on the space $[0, \infty)^n$. The point $\mathbf{r}$ corresponds to the measure $m_{\mathbf{r}}$ on the σ-algebra Σ defined by the formula $m_{\mathbf{r}} = \sum_i r_i \delta_{x_i}$. The measures so drawn and defined are the values of the process (ξ_t) at time t. Moreover, from the formula (2.25) we obtain

$$E_\mu e^{\langle -h, \xi_t \rangle} = E \exp\left(-\sum_{i=1}^n Y_t^i h(x_i)\right) = \prod_{i=1}^n E \exp\left(-Y_t^i h(x_i)\right)$$

$$= \prod_{i=1}^n \exp\left(-\frac{h(x_i) y_i}{1 + h(x_i)\alpha t}\right) = \exp\left(-\sum_{i=1}^n \frac{h(x_i) y_i}{1 + h(x_i)\alpha t}\right).$$

Hence

$$E_\mu e^{\langle -h, \xi_t \rangle} = \exp\left(-\int_{\mathcal{X}} \frac{h(x)\,\mu(dx)}{1 + h(x)\alpha t}\right) \tag{2.29}$$

for $h \in B(\mathcal{X}, \Sigma)$, $h \geq 0$. From the formula

$$LF_h(\mu) = \frac{d}{dt} \exp\left(\int_{\mathcal{X}} \frac{h(x)\,\mu(dx)}{1 - h(x)\alpha t}\right)\bigg|_{t=0}$$

we obtain $LF_h(\mu) = \langle \alpha h^2, \mu \rangle e^{\langle h, \mu \rangle}$ for $h \in B(\mathcal{X}, \Sigma)$ and a measure μ of the form (2.26).

For any measure $\mu \in \mathcal{M}(\mathcal{X})$ there exists a Markov process $(\xi_t)_{t \geq 0}$ with values in the space $\mathcal{M}(\mathcal{X})$ satisfying the initial condition $\xi_0 = \mu$ and Eq. (2.29). In particular, the operator LF_h is of the form (2.23). Equation (2.29) can be written in the form:

$$E_\mu \exp\langle -h, \xi_t \rangle = \exp\langle -u(t, \cdot), \mu \rangle, \tag{2.30}$$

and the function $u(t, x)$ is the solution of the initial problem:

$$\frac{\partial u}{\partial t}(t, x) = -\alpha u^2(t, x), \quad u(0, x) = h(x). \tag{2.31}$$

Let $\mathcal{P}: [0, \infty) \times \mathcal{M}(\mathcal{X}) \times \mathcal{B}(\mathcal{M}(\mathcal{X})) \to [0, 1]$ be the transition probability function for the Markov process $(\xi_t)_{t \geq 0}$. The process $(\xi_t)_{t \geq 0}$ and the function $\mathcal{P}$ have some interesting properties that follow from the formula (2.29):

(a) for a given set $A \in \mathscr{B}(\mathscr{X})$, the Markov process $t \mapsto \xi_t(A)$ has the same distribution as the process Y_t satisfying Eq. (2.24) with the initial condition $Y_0 = \xi_0(A)$,
(b) if supp $\xi_0 \subseteq A$, then supp $\xi_t \subseteq A$ a.s.,
(c) if $A, B \in \mathscr{B}(\mathscr{X})$ and $A \cap B = \emptyset$, then the processes $t \mapsto \xi_t(A)$ and $t \mapsto \xi_t(B)$ are independent,
(d) the distribution of the process $(\xi_t)_{t \geq 0}$ with the initial condition $\mu_1 + \mu_2$, $\mu_1, \mu_2 \in \mathscr{M}(\mathscr{X})$ is the same as the distribution of the sum of two independent copies of the process $(\xi_t)_{t \geq 0}$ satisfying the initial conditions μ_1 and μ_2.

Condition (d) can be written in the form

$$\mathscr{P}(t, \mu_1 + \mu_2, \cdot) = \mathscr{P}(t, \mu_1, \cdot) * \mathscr{P}(t, \mu_2, \cdot), \qquad (2.32)$$

where the convolution $\pi_1 * \pi_2$ of distributions π_1 and π_2 on $\mathscr{M}(\mathscr{X})$ is a new distribution defined by

$$(\pi_1 * \pi_2)(A) = \iint_{\mathscr{M}^2} \mathbf{1}_A(\mu + \nu) \, \pi_1(d\mu) \, \pi_2(d\nu).$$

Chapter 3
Central Limit-Type Theorems

Chapter 2 was devoted to limit passages of sequences of individual-based models to some, usually deterministic, limit. Here we investigate the behaviour of the fluctuation processes, i.e. the difference between the converging process and the limit. We show that after appropriate rescaling the fluctuation process can converge to some Gaussian process. We use this approach to the models from the previous chapter to prove examples of CLT-type theorems. We will start with an introductory example.

3.1 An Introductory Example

We will use the following theorem stated in the setting of Sect. 2.2.

Theorem 3.1 *Let $(X_t^N)_{t\geq 0}$ be a sequence of jump processes like in Theorem 2.1 and let $b^N(\mathbf{x}) \to b(\mathbf{x})$, and $a^N(\mathbf{x}) \to 0$ uniformly, where b is a Lipschitz function. Let moreover $(\alpha_N)_{N\in\mathbb{N}}$ be a sequence of positive numbers converging to infinity and*

(i) $\alpha_N^2 a^N(\mathbf{x}) \to \sigma(\mathbf{x})$ *almost uniformly,*

(ii) $\displaystyle\mathop{\forall}_{r>0} \mathop{\forall}_{\varepsilon>0} \lim_{N\to\infty} \sup_{\{\mathbf{x}:\,\|\mathbf{x}\|\leq r\}} \alpha_N \int \|\mathbf{y}-\mathbf{x}\|^2 \mathbf{1}_{\{\|\mathbf{y}-\mathbf{x}\|>\varepsilon\}}(\mathbf{y}) K^N(\mathbf{x}, d\mathbf{y}) = 0.$

Electronic Supplementary Material The online version of this chapter (https://doi.org/10.1007/978-3-031-75270-4_3) contains supplementary material, which is available to authorized users.

© The Author(s), under exclusive license to Springer Nature Switzerland AG 2024
R. Rudnicki, R. Wieczorek, *Individual-Based Models and Their Limits*,
SpringerBriefs in Mathematical Methods,
https://doi.org/10.1007/978-3-031-75270-4_3

If the sequence X_0^N is weakly convergent to x_0 then the process X_t^N converges to a solution to the differential equation

$$\frac{d\mathbf{x}_t}{dt} = b(\mathbf{x}_t)$$

and the processes

$$Y_t^N = X_t^N - X_0^N - \int_0^t b(X_s^N)\,ds$$

converge to a Gaussian martingale Y_t with $Y_0 = 0$ and with a generator

$$Lg(\mathbf{x}) = \frac{1}{2}\sum_{i,j=1}^n a^{ij}(t)\frac{\partial^2 g(\mathbf{x})}{\partial x^i\,\partial x^j},$$

where $a_{ij}(t) = \sum_k \sigma_{ik}(\mathbf{x}_t)\sigma_{jk}(\mathbf{x}_t)$.

By a *Gaussian martingale* we mean a martingale which has independent Gaussian distributed increments. The proof of this theorem can be found in [59] Ch. IX, Th. 4.26.

Let us now consider a sequence of the birth and death processes with individuals of mass $\frac{1}{N}$, i.e. random walks on $\frac{1}{N}\mathbb{N}$ such that the process jumps from point x to point $x + \frac{1}{N}$ with intensity bNx and to $x - \frac{1}{N}$ with intensity dNx, cf. Sect. 2.2. The kernel is now $K^N(x, \{x - 1/N\}) = dNx$, $K^N(x, \{x + 1/N\}) = bNx$ and zero if none of $x - \frac{1}{N}$ or $x + \frac{1}{N}$ is in B. Then

$$b^N(x) = \frac{1}{N}bNx - \frac{1}{N}dNx = (b-d)x, \quad a^N(x) = \frac{(b+d)x}{N} \xrightarrow{N\to\infty} 0,$$

and by Theorem 2.1, if X_0^N converges weakly to x_0 then the sequence X^N converges to the "deterministic diffusion" x_t (see Fig. 3.1) given by

$$dx_t = (b-d)x_t\,dt, \tag{3.1}$$

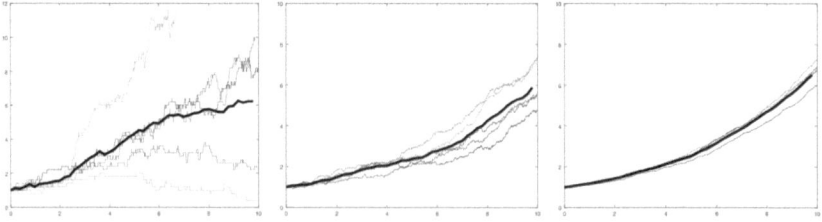

Fig. 3.1 Sample trajectories of birth and death process X^N with $b - d = 0.2$ and $N = 5$, $N = 50$ and $N = 500$. Bold black line is the average

3.1 An Introductory Example

so
$$x_t = x_0 \, e^{(b-d)t}.$$

Now, we state the question, what is the behaviour of the sequence of *fluctuation processes*
$$Y_t^N = \sqrt{N}\left(X_t^N - x_t\right).$$

We will show that if
$$\sqrt{N}\left(X_0^N - x_0\right) \to Y_0, \tag{3.2}$$

then Y_t^N converge in distribution on $\mathbb{D}([0,T], \mathbb{R})$ to the solution Y_t of the stochastic equation
$$dY_t = \lambda Y_t \, dt + \sqrt{\bar{\lambda} x_0 e^{\lambda t}} \, dW_t, \tag{3.3}$$

where $\lambda = b - d$ and $\bar{\lambda} = b + d$ and $(W_t)_{t \geq 0}$ is the standard Brownian motion.

To prove this fact, first, we use Theorem 3.1 to get that
$$M_t^N = \sqrt{N}\left(X_t^N - X_0^N - \int_0^t \lambda X_s^N \, ds\right)$$

converge to the Gaussian martingale
$$M_t = \int_0^t \sqrt{\bar{\lambda} x_0 e^{\lambda s}} \, dW_s.$$

Then, we observe that
$$Y_t^N = \sqrt{N}\left(X_0^N + \int_0^t \lambda X_s^N \, ds + \frac{1}{\sqrt{N}} M_t^N - x_0 - \int_0^t \lambda x_s \, ds\right)$$
$$= \sqrt{N}\left(X_0^N - x_0\right) + \int_0^t \lambda Y_s^N \, ds + M_t^N. \tag{3.4}$$

If $\lambda = b - d = 0$ then we have $Y_t^N = \sqrt{N}\left(X_0^N - x_0\right) + M_t^N$, so the convergence of Y^N to the solution of (3.3) is proven. Note, that in this case (M_t) is just $\sqrt{b+d}$ times the standard Brownian motion (see Fig. 3.2) and if $\sqrt{N}\left(X_0^N - x_0\right) \to 0$ then $Y_t^N \to M_t$. If $\lambda \neq 0$ it looks obvious from (3.4) that having (3.2) processes Y^N converge to the solution of (3.3). To prove it strictly in this simple case, one can use some general facts about diffusion, like e.g. Theorem 4.8 from [59, Ch. IX], but we will do it with a general scheme which will work for more complicated cases

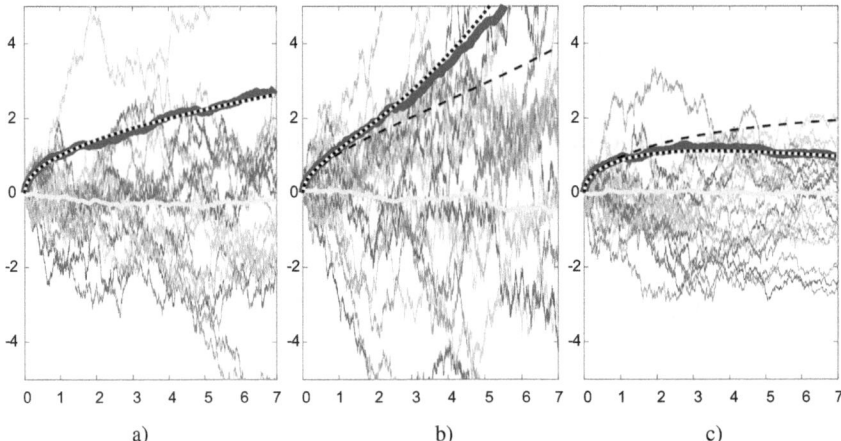

Fig. 3.2 Sample trajectories of the rescaled birth and death process $Y^N = \sqrt{N} X^N$ starting from 0 with (**a**) $b - d = 0$, (**b**) $b - d > 0$, (**c**) $b - d < 0$. The gray bold line is the average, the bold black line is the empirical standard deviation of (Y_t^N) with $N = 500$. The black and white dotted line is the square root of theoretical variance, for (**a**) $\mathrm{E}\,(Y_t)^2 = x_0(b+d)t$, for (**b**) and (**c**) $\mathrm{E}\,(Y_t)^2 = x_0 \frac{b+d}{b-d} e^{(b-d)t} \left(e^{(b-d)t} - 1\right)$. The dashed line in (**b**) and (**c**) is the square root of variance of M_t, where $\mathrm{E}(M_t)^2 = x_0 \frac{b+d}{b-d}\left(e^{(b-d)t} - 1\right)$

later. Namely, the procedure is to prove tightness of $(Y^N)_{N \in \mathbb{N}}$ and then identify the limit. For the proof of tightness we need to check condition (3.8) and the Aldous criterion (3.10).

In a similar manner as in the proof of Lemma 3.1 in the next section, we get that $\mathrm{E}\, X_t^N = x_0\, e^{(b-d)t}$ (cf. Fig. 3.1) and $\mathrm{E}(M_t^N)^2 = \frac{b+d}{b-d} x_0 \left(e^{(b-d)t} - 1\right)$ (cf. Fig. 3.2). So, by Doob's inequality,

$$\mathrm{E}\left[\sup_{0 < t < T} (M_t^N)^2\right] \leq 4 \frac{b+d}{b-d} x_0 \left(e^{(b-d)T} - 1\right).$$

Now, from (3.4) we have

$$\mathrm{E}\left[\sup_{0 \leq t \leq T} (Y_t^N)^2\right] \leq$$
$$3\,\mathrm{E}\left[N\left(X_0^N - x_0\right)^2\right] + 3 \int_0^t \lambda\, \mathrm{E}\left[\sup_{0 \leq t \leq T} (Y_s^N)^2\right] ds + 3\,\mathrm{E}\left[\sup_{0 < t < T} (M_t^N)^2\right].$$

The first and the last term on the right-hand side are bounded uniformly in N, so by Grönwall's inequality also $\mathrm{E}\left[\sup_{0 \leq t \leq T}(Y_t^N)^2\right]$ is bounded. So by Markov's

3.1 An Introductory Example

inequality we get (3.8). Then, since for any stopping time $\tau \leq T$ and $0 < s < \gamma$ we have

$$Y_{\tau+s}^N - Y_\tau^N = \int_\tau^{\tau+s} \lambda Y_t^N dt + M_{\tau+s} - M_\tau,$$

we can get $\mathrm{E}\left[Y_{\tau+s}^N - Y_\tau^N\right] \leq Cs$ for some constant C so by Markov's inequality we obtain the Aldous criterion (3.10) which completes the proof of tightness. Now, we know that every subsequence of Y^N has an accumulation point. We will show that every accumulation point of Y^N satisfies (3.3). Let $(Y_t^{N_n})$ be a subsequence of (Y_t^N) convergent in distribution to some (Y_t). Then, by (3.4) we have

$$Y_t - Y_0 - \int_0^t \lambda Y_s \, ds - \int_0^t \sqrt{\bar{\lambda} x_0 e^{\lambda s}} \, dW_s$$
$$= \left[\sqrt{N_n}\left(X_0^{N_n} - x_0\right) - Y_0\right] + (Y_t - Y_t^{N_n}) + \int_0^t \lambda(Y_s^{N_n} - Y_s) ds + (M_t^{N_n} - M_t).$$

The first term on the right-hand-side converges to 0 by (3.2), the second and third terms by assumption $Y_t^{N_n} \to Y_t$, and we have already observed that $M_t^N \to M_t$. Therefore, the right hand side converges to 0.

By the way, in [59] there is an incorrect Corollary 4.28 that says wrongly that Y_t^N converge to M_t.

Note that we can characterize the birth and death process X^N differently: let $\left(\tilde{X}^i\right)_{i \in \mathbb{N}}$ be a sequence of independent copies of X^1 i.e. continuous birth and death processes with birth rate b and death rate d. Then $\frac{1}{N} \sum_{i=1}^K \tilde{X}^i = x_0 \frac{1}{K} \sum_{i=1}^K \tilde{X}^i$ has the same distribution as the process X^N starting from $x_0 = \frac{K}{N}$. One can easily check that $\mathrm{E} X_t^N = x_0 e^{\lambda t}$ with $\lambda = b - d$, which is the solution of (3.1). So for a fixed time t the convergence of X_t^N to x_t is just the law of large numbers. Similarly, we can write

$$Y_t^N = \sqrt{N} \left(\frac{1}{N} \sum_{i=1}^K \tilde{X}^i - x_0 e^{\lambda t}\right) = \sqrt{x_0} \sqrt{K} \left(\frac{1}{K} \sum_{i=1}^K \tilde{X}^i - e^{\lambda t}\right)$$

and knowing that $\mathrm{E}\left(Y_t^N\right)^2 = \sigma^2(t) = \frac{\bar{\lambda} x_0}{\lambda} e^{\lambda t} \left(e^{\lambda t} - 1\right)$ we can use the central limit theorem to prove that the distribution of Y_t^N converges to the Gaussian distribution with mean 0 and variance $\sigma^2(t)$. Note also that the process Y_t given by (3.3) is Gaussian (but not a martingale) with mean 0 and variance $\sigma^2(t)$, so the convergence of the fluctuation processes Y^N to Y is a functional CLT.

The next two sections provide some tools used in the proofs of CLT-type theorems.

3.2 Poisson Random Measures, Jump Processes and Itô Formula

Let $\mathscr{M}_\sigma$ be a space of all σ-finite Borel measures on a locally compact Polish space $(E, \mathscr{B}(E))$ with a σ-algebra $\mathscr{B}_{\mathscr{M}_\sigma}$ generated by all mappings $\mu \mapsto \mu(B)$ with $B \subset E$ bounded. A random element $\mathfrak{N} : \Omega \to \mathscr{M}_\sigma$ is called a *random measure*. We say that $\mathfrak{N}$ is a *random point measure* (or *integer-valued measure*) if $\mathfrak{N}(B) \in \mathbb{N} \cup \{\infty\}$ for any $B \in \mathscr{B}(E)$.

Let $\tilde{\nu}$ be a σ-finite measure on $(E, \mathscr{B}(E))$. A *Poisson random measure* on $(E, \mathscr{B}(E))$ with intensity $\tilde{\nu}$ is a random point measure such that

(a) for any $B \in \mathscr{B}(E)$ the random variable $\mathfrak{N}(B)$ has the Poisson distribution with parameter $\tilde{\nu}(B)$,
(b) if $B_1, \ldots, B_n$ are disjoint sets from $\mathscr{B}(E)$ then $\mathfrak{N}(B_1), \ldots, \mathfrak{N}(B_n)$ are independent random variables.

Usually, we are interested in time-space Poisson random measures, i.e. Poisson random measures on $(\mathbb{R}_+ \times E, \mathscr{B}(\mathbb{R}_+) \otimes \mathscr{B}(E))$, where $\tilde{\nu}(dt \times dx) = dt\, \nu(dx)$. If $\mathfrak{N}$ is a Poisson random measure with intensity $dt\, \nu(dx)$ and $B \in \mathscr{B}(E)$ is such that $\nu(B) < \infty$, then one can express $\mathfrak{N}$ on $[a, b] \times B$ as

$$\mathfrak{N}\big|_{[a,b]\times B} = \sum_{i=1}^{N} \delta_{t_i, x_i},$$

where $N = \mathfrak{N}([a, b] \times B)$ is a Poisson random variable with parameter $(b-a)\nu(B)$ and (t_i, x_i) are independent random elements on $[a, b] \times B$ with distribution $\frac{dt\, \nu(dx)}{(b-a)\nu(B)}$. Moreover, for any $\mathscr{B}(\mathbb{R}_+) \otimes \mathscr{B}(E)$-measurable function h we have

$$\int_a^b \int_B h(t, x) \mathfrak{N}(dt \times dx) = \sum_{i=1}^{N} h(t_i, x_i).$$

Let us now consider a filtration $(\mathscr{F}_t)_{t \in T}$ on $(\Omega, \mathscr{F}, P)$ satisfying the usual conditions and a time-space Poisson random measure and assume that

- the space E is a subset of $\mathbb{R}$ and the compensating measure ν is finite on bounded sets,
- for any $B \in \mathscr{B}(E)$ and $t \geq 0$ the random variable $\mathfrak{N}([0, t] \times B)$ is $\mathscr{F}_t$-measurable,
- for any $t \geq 0$ the family of random variables $\{\mathfrak{N}((t, t+s) \times B) : B \in \mathscr{B}(E), s > 0\}$ is independent of $\mathscr{F}_t$.
- $h : \Omega \times \mathbb{R}_+ \times E \to \mathbb{R}$ is a process such that for any $t > 0$ the function $h(\cdot, t, \cdot)$ is $\mathscr{F}_t \otimes \mathscr{B}(E)$-measurable and $\int_0^t \int_E h(s, x) \nu(dx) ds$ is a.s. finite.

3.2 Poisson Random Measures, Jump Processes and Itô Formula

Then the following compensator formula holds

$$E \int_0^t \int_E h(s,z) \mathfrak{N}(ds,dz) = E \int_0^t \int_E h(s,z) \nu(dz) ds. \tag{3.5}$$

We define a *compensated Poisson random measure* as $\tilde{\mathfrak{N}}(dt,dz) = \mathfrak{N}(dt,dz) - \nu(dz)dt$. If a process $X(t)$ can be represented as

$$X(t) = X(0) + \int_0^t \beta(s) ds + \int_0^t \sigma(s) dW_s + \int_0^t \int_E h(s,z) \tilde{\mathfrak{N}}(ds,dz) \tag{3.6}$$

then the following generalized Itô formula (called also the Itô-Lévy formula, see [22]) holds

Theorem 3.2 *If $f : \mathbb{R}_+ \times \mathbb{R} \to \mathbb{R}$ is a twice differentiable function then the process $Y(t) = f(t, X(t))$ satisfies*

$$\begin{aligned}
dY(t) =& \frac{\partial f}{\partial t}(t, X(t)) dt + \frac{\partial f}{\partial x}(t, X(t)) \beta(t) dt + \frac{\partial f}{\partial x}(t, X(t)) \sigma(t) dW_t \\
&+ \frac{1}{2} \frac{\partial^2 f}{\partial x^2}(t, X(t)) \sigma^2(t) dt \\
&+ \int_E \left[f(t, X(t) + h(t,z)) - f(t, X(t)) - \frac{\partial f}{\partial x}(t, X(t)) h(t,z) \right] \nu(dz) dt \\
&+ \int_E \left[f(t^-, X(t^-) + h(t^-, z)) - f(t^-, X(t^-)) \right] \tilde{\mathfrak{N}}(dt, dz).
\end{aligned}$$

The proof can be found, e.g., in [51] or in a more general case in [59].

We will give a simple example using the above theory to prove a lemma that is used several times in the book.

Example 3.1 (A Simple Birth Process) A simple birth process (sometimes called the Yule process) is an integer-valued process that jumps always by one, with a rate proportional to current value. Let us consider a little more general process. Let X_t be a jump process on $\mathbb{R}_+$ such that jumps by $\frac{1}{N}$ with rate bNX_t, so its generator is $Af(x) = bNx(f(x + \frac{1}{N}) - f(x))$. We can write it using the following equation with respect to the Poisson random measure

$$dX_t = \int_{\mathbb{R}_+} \frac{1}{N} 1_{[0, bNX_{t^-}]}(z) \mathfrak{N}(dt, dz), \tag{3.7}$$

where $\mathfrak{N}$ is a Poisson random measure on $\mathbb{R}^+ \times \mathbb{R}^+$ with Lebesgue measure on $\mathbb{R}_+$ as the intensity measure ν.

Lemma 3.1 *Let X_t be a rescaled Yule process as above and $\mathrm{E}\,X_0^2 < \infty$. Then $\mathrm{E}\,X_t = \mathrm{E}\,X_0 e^{bt}$ and $\mathrm{E}[X_t^2] = \frac{1}{N} e^{bt}(e^{bt} - 1) + (\mathrm{E}\,X_0)^2 e^{2bt}$.*

Proof Using the compensator formula (3.5) to (3.7) we obtain the equation for $\mathrm{E}\,X_t$

$$\frac{d\,\mathrm{E}\,X_t}{dt} = b\,\mathrm{E}\,X_t,$$

which gives the formula for $\mathrm{E}\,X_t$. Then, we write X_t in the form (3.6)

$$dX_t = b\,X_t dt + \int_{\mathbb{R}_+} \frac{1}{N} 1_{[0,bNX_{t^-}]}(z)\,\tilde{\mathfrak{N}}(dt, dz),$$

and use the Itô-Lévy formula (Theorem 3.2) for $f(t, x) = x^2$ to get

$$\begin{aligned}
dX_t^2 =\,& 2b\,X_t^2\,dt + \int_{\mathbb{R}_+} \frac{1}{N^2} 1_{[0,bNX_{t^-}]}(z)\nu(dz)dt \\
&+ \int_{\mathbb{R}_+} \left[\frac{2}{N} X_{t^-} + \frac{1}{N^2}\right] 1_{[0,bNX_{t^-}]}(z)\mathfrak{N}(dt, dz) \\
&- \int_{\mathbb{R}_+} \left[\frac{2}{N} X_{t^-} + \frac{1}{N^2}\right] 1_{[0,bNX_{t^-}]}(z)\,\nu(dz)dt \\
=\,& 2b\,X_t^2\,dt + \int_{\mathbb{R}_+} \left[\frac{2}{N} X_{t^-} + \frac{1}{N^2}\right] 1_{[0,bNX_{t^-}]}(z)\,\mathfrak{N}(dt, dz) \\
&- \int_{\mathbb{R}_+} \frac{2}{N} X_{t^-}\,\nu(dz)dt.
\end{aligned}$$

Now, using again the compensator formula (3.5) we get

$$\frac{d\,\mathrm{E}[X_t^2]}{dt} = 2b\,\mathrm{E}[X_t^2] + \frac{1}{N} b\,\mathrm{E}\,X_t.$$

and solving the equation for $\mathrm{E}[X_t^2]$ we obtain $\mathrm{E}[X_t^2] = \frac{1}{N} e^{bt}(e^{bt} - 1) + (\mathrm{E}\,X_t)^2$. □

3.3 Convergence of Càdlàg Processes

When studying stochastic processes with jumps, one introduces a space of càdlàg (i.e. *right continuous with left limits*) functions $\mathbb{D}([0, T], X)$ in which the functions are defined on the interval $[0, T]$ and take values in a Polish space (X, d). In the Skorohod space $\mathbb{D} = \mathbb{D}([0, T], X)$ one can introduce the so called *Skorohod topology* J_1, which is metrizable in a complete way by the *Skorohod metric* d_0 (see Section B.3). If a sequence of càdlàg functions converges to a continuous function, then the convergence in the Skorohod metric d_0 coincides with the uniform convergence on compact sets. By introducing the Skorohod space, we can look at a stochastic process as a random variable with values in the sample path space and define the *convergence in the sense of distribution* of a sequence of processes $(\xi_t^n)_{t\geq 0}$ to a process $(\xi_t)_{t\geq 0}$ as a weak convergence of distributions of $(\xi_t)_{t\geq 0}$ in the space of measures.

We will now give necessary and sufficient conditions for the tightness of measures on the space $\mathbb{D}$. For this purpose we define a *càdlàg modulus*:

$$\omega_f^c(\delta) := \inf_{\Pi} \max_{1\leq i\leq k} \omega_f([t_{i-1}, t_i)),$$

where $\omega_f(F) := \sup_{s,t\in F} d(f(s), f(t))$, where the infimum runs over all partitions $\Pi = \{0 = t_0 < t_1 < \cdots < t_k = T\}$ with $\min_i\{t_i - t_{i-1}\} > \delta$. It can be shown that a function f is càdlàg if and only if $\lim_{\delta\to 0}\omega_f^c(\delta) = 0$. A sequence of probability measures (P_n) on $\mathbb{D}$ is tight if and only if the following two conditions are satisfied:

for all $t \in [0, T]$, $\varepsilon > 0$ there is a compact set $K_{t,\varepsilon}$ s.t. $P_n\left(f(t) \notin K_{t,\varepsilon}\right) \leq \varepsilon$, (3.8)

$$\lim_{\delta\to 0} \limsup_{n\to\infty} P_n(\{f \in \mathbb{D}: \omega_f^c(\delta) \geq \varepsilon\}) = 0 \quad \text{for all } \varepsilon > 0. \tag{3.9}$$

The following useful sufficient condition for tightness is called the Aldous criterion. If the sequence of probability measures P_n on $\mathbb{D}$ satisfies

$$\lim_{\gamma\to 0}\limsup_{n_0\to\infty} \sup_{\tau\in\mathcal{T}_T, s<\gamma} P_n\left(d(f(\tau+s), f(\tau)) > \varepsilon\right) = 0, \tag{3.10}$$

where $\mathcal{T}_T$ is the set of all stopping times bounded by T, then condition (3.9) is satisfied.

3.4 CLT-Type Theorems for IBM

Chapter 2 was devoted to limit passages from individual-based models to macroscopic models. The IBMs were described as a sequence of measure-valued jump processes $(\xi^N)_{t\geq 0}$ and the limit was usually a deterministic evolution of measure, say $(\xi_t)_{t\geq 0}$. Analogously to the introductory example above, an interesting question arises, what is the behaviour of the fluctuations $\xi^N - \xi$. While the values of the processes ξ^N and ξ are positive measures, their difference is signed-measured-valued. The problem appears, that the space of signed measures with the topology of weak convergence, which is natural here, is in general not metrizable by a complete metric, so we need to consider the fluctuation process in some bigger space e.g. the space of Schwartz distributions. Nevertheless, it turns out that the rescaled fluctuations $\alpha_N(\xi^N - \xi)$ may converge to some Gaussian process in a properly chosen space. The usual rescaling is $\alpha_N = \sqrt{N}$, but sometimes a different one may be appropriate. In this section we show some examples of such CLT-type results.

In Sect. 3.1 we have seen a sequence of real jump processes X_t^N which converges to a deterministic process $x_t = x_0 e^{\lambda t}$, and we observed that the fluctuation processes $Y_t^N = \sqrt{N}\left(X_t^N - x_t\right)$ converge to a Gaussian process Y_t satisfying Eq. (3.3). Note that the measure-valued process ξ^N in Example 2.5, which jumps from the measure $\mu = \frac{1}{N}\sum_i \delta_{x_i}$ to the measure $\mu + \frac{1}{N}\delta_{x_i}$ with intensity $b(x_i)$ and to the measure $\mu - \frac{1}{N}\delta_{x_i}$ with intensity $d(x_i)$, is a spatial version of X^N. Namely, for every phenotype x_i there is an independent birth and death process like X^N with $b = b(x_i)$ and $d = d(x_i)$. It is natural to expect, that the fluctuation processes

$$\eta_t^N = \sqrt{N}\left(\xi_t^N - \xi_t\right) \tag{3.11}$$

should satisfy a central limit theorem similar to that in Sect. 3.1. Actually, if the initial processes ξ_0^N and ξ_0 are concentrated on finite number of points, then we just have finite number of processes like Y^N above. But usually, the values of processes η_t^N are extremely irregular signed measures and we cannot expect the weak limit η_t to be a signed measure, but it will take values in some space of distributions.

Note, that since the generator of ξ_N is

$$L_N F(\mu) = \sum_i \left(b(x_i)\left(F(\mu + \tfrac{1}{N}\delta_{x_i}) - F(\mu)\right) + d(x_i)\left(F(\mu - \tfrac{1}{N}\delta_{x_i}) - F(\mu)\right)\right),$$

for $F\colon \mathcal{N}_N \to \mathbb{R}$, for any measurable and bounded function $h\colon \mathcal{X} \to \mathbb{R}$ we have

$$\langle h, \xi_t^N\rangle = \langle h, \xi_0^N\rangle + \int_0^t \left\langle (b-d)h, \xi_s^N\right\rangle ds + M_t^{h,N}, \tag{3.12}$$

where $M_t^{h,N}$ is a martingale with the predictable quadratic variation

$$\left\langle M^{h,N}\right\rangle_t = \int_0^t \frac{1}{N}\left\langle (b-d)h^2, \xi_s^N\right\rangle ds \tag{3.13}$$

(see Lemma 3.3 below for the proof of this fact). Since the macroscopic limit of ξ^N is $\xi_t(dx) = \Pi_t \xi_0(dx) = e^{(b(x)-d(x))t}\xi_0(dx)$, we have

$$\langle h, \xi_t\rangle = \langle h, \xi_0\rangle + \int_0^t \langle (b-d)h, \xi_s\rangle\, ds$$

3.4 CLT-Type Theorems for IBM

and thereby the fluctuation processes satisfy

$$\langle h, \eta_t^N \rangle = \langle h, \eta_0^N \rangle + \int_0^t \langle (b-d)h, \eta_s^N \rangle ds + \tilde{M}_t^{h,N}, \tag{3.14}$$

where $\tilde{M}_t^{h,N} = \sqrt{N} M_t^{h,N}$ and $\langle \tilde{M}^{h,N} \rangle_t = \int_0^t \langle (b-d)h^2, \xi_s^N \rangle ds$.

To check that the real-valued martingales $(\tilde{M}_t^{h,N})$ converge to a martingale we can use the following martingale central limit theorem

Theorem 3.3 *Let (M_t) be a Gaussian martingale with $M_0 = 0$ and with generator*

$$Lg(\mathbf{x}) = \frac{1}{2} \sum_{i,j=1}^{n} a^{ij}(t) \frac{\partial^2 g(\mathbf{x})}{\partial x^i \partial x^j},$$

where a is nonnegative definite. If (M_t^N) are local martingales with uniformly bounded jumps and

(i) $\lim_{N \to \infty} E[\sup_{t \in [0,T]} |M_t^N - M_{t-}^N|] = 0,$

(ii) *predictable quadratic variations $\langle M^N \rangle_t$ converge in distribution to $C(t) = \int_0^t a(s) ds$,*

then (M_t^N) converge in distribution on $\mathbb{D}([0, T], \mathbb{R}^d)$ to (M_t).

The proof can be found in [59, Ch. VIII], Theorem 3.11 or [42, Ch. 7], Theorem 1.4.

Jumps of $\langle h, \xi_t^N \rangle$ are bounded by $\frac{1}{N} \|h\|_\infty$. Since (ξ_s^N) converge in distribution to (ξ_s), the predictable quadratic variations $\langle \tilde{M}^{h,N} \rangle_t$ converge to

$$\int_0^t \langle (b-d)h^2, \xi_s \rangle ds. \tag{3.15}$$

Therefore, by Theorem 3.3, martingales $(\tilde{M}_t^{h,N})$ converge to a martingale $(\tilde{M}_t^h)$ with the predictable quadratic variation given by (3.15).

For this reason, by (3.14) one can expect that (η_t^N) converge to a process (η_t) such that

$$\langle h, \eta_t \rangle = \langle h, \eta_0 \rangle + \int_0^t \langle (b-d)h, \eta_s \rangle ds + \tilde{M}_t^h$$

for any measurable and bounded function $h: \mathscr{X} \to \mathbb{R}$. But this requires some careful consideration about the spaces in which processes (η_t) can be defined, to make the convergence precise. In this very simple case, where the evolution is independent for every $x \in \mathscr{X}$, we have rather wide choice of spaces. One can work on the space of Schwartz distributions like e.g. in [57] or one can use some Sobolev spaces as in [77, 90] or other Hilbert space e.g. [44]. An interesting possibility is the space of tempered (Schwartz) distributions $\mathscr{S}'$, where $\mathscr{S} = \mathscr{S}(\mathscr{X})$ is the Schwartz space of functions with rapidly decreasing derivatives, which is known to be a nuclear space. In this space the proof of tightness of the fluctuation processes is relatively simple thanks to Theorem 3.1 in [81], which says that a process η_t in $\mathscr{S}'$ is tight, if for every $f \in \mathscr{S}$ the process $\langle \eta_t, f \rangle$ is tight. Using of the space $\mathscr{S}'$ has two drawbacks. One that this space is unreasonably big extension of the space of measures. And second, technical one, that it is not a Banach space and working with the family of seminorms that define topology is less comfortable.

Let $\mathscr{H}^k$ denote the Sobolev space of $L^2(\mathscr{X})$ functions with weak derivatives up to order k in $L^2(\mathscr{X})$, and $\mathscr{H}^{-k}$ its dual space—the space of continuous linear functionals on $\mathscr{H}^k$. Assume that $\mathscr{X} \subset \mathbb{R}^d$ is bounded with C^1 boundary. Let $\bar{d} = \lceil \frac{d}{2} \rceil$. We will prove the following.

Theorem 3.4 *Assume that b and d are $C_b^{\bar{d}+1}$ functions. Let $\sqrt{N}(\xi_0^N - \xi_t)$ converge weakly in $\mathscr{H}^{-(\bar{d}+1)}$ to some random variable η_0 and assume that ξ_0 is a finite measure, $\sup_N \mathrm{E}\langle 1, \xi_0^N \rangle < \infty$, and $\sup_N \mathrm{E} \|\eta_0^N\|_{TV} < \infty$. The fluctuation processes (η_t^N) (3.11) converge in distribution on the space $\mathbb{D}([0, T], \mathscr{H}^{-(\bar{d}+1)})$ to a process (η_t) described by the equation*

$$\eta_t = \eta_0 + \int_0^t (b - d)\eta_s \, ds + \tilde{M}_t, \qquad (3.16)$$

where $\tilde{M}_t$ is a $C_b(\mathscr{X})'$-valued centered Gaussian martingale with a quadratic covariation

$$\left\langle \langle g, \tilde{M}_t \rangle, \langle h, \tilde{M}_t \rangle \right\rangle = \int_0^t \langle (b-d)gh, \xi_s \rangle \, ds, \qquad (3.17)$$

and the integral in (3.16) is a Bochner integral.

The proof of this theorem follows similarly to the proof for the introductory example in Sect. 3.1, namely by proving the tightness of (η_t^N) and the identification of the limit as the unique solution to (3.16). The difference is that we need to define the fluctuation processes as $\mathbb{D}([0, T], \mathscr{H}^{-(\bar{d}+1)})$-valued and show the tightness there. We will divide the proof into several lemmas.

First, note that

Lemma 3.2 $\mathrm{E}\left[\sup_{0 \leq s \leq t}\langle 1, \xi_t^N \rangle\right]$ *and* $\mathrm{E}\left[\sup_{0 \leq s \leq t}\langle 1, \xi_t^N \rangle^2\right]$ *are bounded uniformly in N.*

This lemma follows from the fact that the total mass of particles $\langle 1, \xi_t^N \rangle$ is bounded in distribution by a simple pure birth process that jumps by $\frac{1}{N}$ with rate $\bar{b} = \sup_{x \in \mathscr{X}} b(x)$, like in Example 3.1. The first and second moment of this process is given in Lemma 3.1.

Lemma 3.3 *Process $(M_t^{h,N})$ defined in (3.12) is a martingale with the predictable quadratic variation (3.13).*

3.4 CLT-Type Theorems for IBM

Proof We need to write $\langle h, \xi_t^N \rangle$ as a stochastic integral with respect to Poisson random measures. To do so, we need to give to each particle its own Poisson clock, and therefore we need to distinguish the particles. One way to do it (cf. [49]), is introducing an arbitrary order on $\mathscr{X}$, say $\preccurlyeq$, and projecting the space of measures $\mathscr{N}_N$ in the space $(\mathscr{X} \cup \{*\})^{\mathbb{N}}$ by defining a function

$$H\left(\frac{1}{N}\sum_{i=1}^{n}\delta_{x_i}\right) = (x_{\sigma(1)}, \ldots, x_{\sigma(n)}, *, *, \ldots),$$

where σ is a permutation of $\{1, \ldots, n\}$ such that $x_{\sigma(1)} \preccurlyeq \cdots \preccurlyeq x_{\sigma(n)}$. Then we take two sequences of Poisson random measures $\mathfrak{N}_b^i(ds, dz)$ and $\mathfrak{N}_d^i(ds, dz)$ on $[0, \infty) \times [0, \infty)$, for the birth and death, respectively, and we can write

$$\langle h, \xi_t^N \rangle = \langle h, \xi_0^N \rangle + \int_0^t \int_{\mathbb{R}_+} \frac{1}{N} \sum_{i=1}^{\langle 1, N\xi_{s-}^N \rangle} h\left(H_i(\xi_{s-}^N)\right) 1_{[0, b(H_i(\xi_{s-}^N))]}(z)\, \mathfrak{N}_b^i(ds, dz)$$

$$- \int_0^t \int_{\mathbb{R}_+} \frac{1}{N} \sum_{i=1}^{\langle 1, N\xi_{s-}^N \rangle} h\left(H_i(\xi_{s-}^N)\right) 1_{[0, d(H_i(\xi_{s-}^N))]}(z)\, \mathfrak{N}_d^i(ds, dz)$$

and thus

$$M_t^{h,N} = \int_0^t \int_{\mathbb{R}_+} \frac{1}{N} \sum_{i=1}^{\langle 1, N\xi_{s-}^N \rangle} h\left(H_i(\xi_{s-}^N)\right) 1_{[0, b(H_i(\xi_{s-}^N))]}(z)\, \tilde{\mathfrak{N}}_b^i(ds, dz)$$

$$- \int_0^t \int_{\mathbb{R}_+} \frac{1}{N} \sum_{i=1}^{\langle 1, N\xi_{s-}^N \rangle} h\left(H_i(\xi_{s-}^N)\right) 1_{[0, d(H_i(\xi_{s-}^N))]}(z)\, \tilde{\mathfrak{N}}_d^i(ds, dz),$$

where $\tilde{\mathfrak{N}}_b^i$ and $\tilde{\mathfrak{N}}_d^i$ are compensated Poisson measures (see Sect. 3.2). Since the number of particles $\langle 1, N\xi_t^N \rangle$ is less than a pure birth process (Yule process) with birth rate $\sup_{x \in \mathscr{X}} b(x)$ (b is bounded), all sums over i are finite with probability one in the formula above. Therefore, by the compensation formula (3.5), $M_t^{h,N}$ is a martingale and by Itô-Lévy formula

$$\left(M_t^{h,N}\right)^2 = \int_0^t \int_{\mathbb{R}_+} \frac{2}{N} M_{s-}^{h,N} \sum_{i=1}^{\langle 1, N\xi_{s-}^N \rangle} h\left(H_i(\xi_{s-}^N)\right) 1_{[0, b(H_i(\xi_{s-}^N))]}(z)\, \tilde{\mathfrak{N}}_b^i(ds, dz)$$

$$- \int_0^t \int_{\mathbb{R}_+} \frac{2}{N} M_{s-}^{h,N} \sum_{i=1}^{\langle 1, N\xi_{s-}^N \rangle} h\left(H_i(\xi_{s-}^N)\right) 1_{[0, d(H_i(\xi_{s-}^N))]}(z)\, \tilde{\mathfrak{N}}_d^i(ds, dz)$$

$$+ \int_0^t \int_{\mathbb{R}_+} \frac{1}{N^2} \sum_{i=1}^{\langle 1, N\xi_{s-}^N\rangle} h^2\left(H_i(\xi_{s-}^N)\right) 1_{[0,b(H_i(\xi_{s-}^N))]}(z)\, \mathfrak{N}_b^i(ds, dz)$$

$$+ \int_0^t \int_{\mathbb{R}_+} \frac{1}{N^2} \sum_{i=1}^{\langle 1, N\xi_{s-}^N\rangle} h^2\left(H_i(\xi_{s-}^N)\right) 1_{[0,d(H_i(\xi_{s-}^N))]}(z)\, \mathfrak{N}_d^i(ds, dz)$$

Again, by the compensation formula we can see that subtracting (3.13) from $(M_t^{h,N})^2$ we get a martingale, so (3.13) is really the predictable quadratic variation of $M_t^{h,N}$. □

We have defined (η_t^N) by (3.11) as a difference of measure-valued processes, so as a singed-measure-valued process. Similarly, we could define $\tilde{M}_t^N = \sqrt{N}(\xi_t^N - \xi_0^N - \int_0^t (b-d)\xi_s^N ds)$ as signed-measure-valued. But we need them to be $\mathscr{H}^{-\bar{d}}$-valued, so we define

$$\tilde{M}_t^N h = \tilde{M}_t^{h,N} = \sqrt{N}\left(\langle h, \xi_t^N\rangle - \langle h, \xi_0^N\rangle - \int_0^t \langle h(b-d), \xi_s^N\rangle ds\right), \quad (3.18)$$

for $h \in \mathscr{H}^{\bar{d}}$. Now, we check that

Lemma 3.4 *Process $\tilde{M}_t^N$ is a square-intregrable $\mathscr{H}^{-\bar{d}}$-valued càdlàg martingale with the predictable quadratic covariation*

$$\langle \tilde{M}_t^N(g), \tilde{M}_t^N(h)\rangle_t = \int_0^t \langle (b-d)gh, \xi_s^N\rangle ds. \quad (3.19)$$

Proof By Sobolev imbedding theorem, $\mathscr{H}^{\bar{d}}$ is continuously included in $C_b(\mathscr{X})$, so we can take h to be bounded continuous and (3.18) is well defined. Let $(\varphi_k)_{k\in\mathbb{N}}$ be an orthonormal base of $\mathscr{H}^{\bar{d}}$ consisting of bounded continuous functions. Then by the Parseval formula

$$\|\eta\|_{\mathscr{H}^{-\bar{d}}}^2 = \sum_{k=1}^{\infty} |\langle \varphi_k, \eta\rangle|^2$$

and since $\delta_x \in \mathscr{H}^{-\bar{d}}$, for any $x \in \mathscr{X}$

$$\sum_{k=1}^{\infty} \varphi_k^2(x) = \|\delta_x\|^2 = \left(\sup_{\|\varphi\|_{\mathscr{H}^{\bar{d}}}\leq 1} |\varphi(x)|\right)^2 \leq \left(\sup_{\|\varphi\|_{\mathscr{H}^{\bar{d}}}\leq 1} \|\varphi\|_{C_b(\mathscr{X})}\right)^2 \leq C$$

3.4 CLT-Type Theorems for IBM

by the Sobolev imbedding theorem. Then, by Doob's inequality we have

$$\mathrm{E}\left[\sum_{k=1}^{\infty} \sup_{t\in[0,T]} \left|\tilde{M}_t^{\varphi_k,N}\right|^2\right] \leq 4\,\mathrm{E}\left[\sum_{k=1}^{\infty}\left|\tilde{M}_T^{\varphi_k,N}\right|^2\right] = 4\sum_{k=1}^{\infty}\mathrm{E}\left[\langle\tilde{M}^{\varphi_k,N}\rangle_T\right] \quad (3.20)$$

$$= 4\,\mathrm{E}\left[\langle\tilde{M}^{\sum_{k=1}^{\infty}\varphi_k,N}\rangle_T\right] \leq 4C\,\mathrm{E}\left[\langle\tilde{M}^{1,N}\rangle_T\right]$$

$$= 4CT(\|b\|_\infty + \|d\|_\infty)\,\mathrm{E}\left[\sup_{t\in[0,T]}\langle 1,\xi_t^N\rangle\right]$$

which is bounded uniformly in N by Lemma 3.2. Now again by the Parseval identity $\mathrm{E}\left[\sup_{t\in[0,T]}\|\tilde{M}_t^N\|^2_{\mathscr{H}^{-\bar{d}}}\right]$ is less than (3.20), thus $\tilde{M}_t^N$ is square integrable. To prove that $\tilde{M}_t^N$ is càdlàg, by (3.20), it suffices to check the continuity condition on a finite number of basis functions φ_k, so the right hand continuity and existence of left-hand limits follows from every $\tilde{M}_t^{\varphi_k,N}$ being càdlàg. The martingale condition for $\tilde{M}_t^N$ (in the Bochner integral sens) also follows from the martingale property of all $\tilde{M}_t^N(\varphi) = \tilde{M}_t^{\varphi,N}$. By Lemma 3.3 and the polarization formula (A.8), we have (3.19). □

Lemma 3.5 *We can understand η_t^N as an $\mathscr{H}^{-\bar{d}}$-valued process, by assuming $\eta_t^N h = \langle h,\eta_t^N\rangle$ for $h\in\mathscr{H}^{\bar{d}}$. Moreover η^N is càdlàg and*

$$\eta_t^N = \eta_0^N + \int_0^t (b-d)\eta_s^N ds + \tilde{M}_t^N, \quad (3.21)$$

where the integral is a Bochner integral.

The proof follows from the fact that $\mathscr{H}^{\bar{d}}$ is imbedded into $C_b(\mathscr{X})$, so $\langle h,\eta_t^N\rangle$ is well defined for $h\in\mathscr{H}^{\bar{d}}$ and $(b-d)\eta_s^N$ is bounded, so Bochner integrable in $\mathscr{H}^{\bar{d}}$.

Lemma 3.6 *The sequence $\tilde{M}^N$ is tight in $\mathbb{D}([0,T],\mathscr{H}^{-(\bar{d}+1)})$ and its only limit point is a $\mathscr{H}^{-(\bar{d}+1)}$-valued Gaussian martingale with the predictable covariation given by* (3.17).

Proof The proof is based on the Aldous criterion (see Sect. 3.3). The compact imbedding of $\mathscr{H}^{-\bar{d}}$ into $\mathscr{H}^{-(\bar{d}+1)}$ and uniform boundedness of $\tilde{M}^N$ in $\mathscr{H}^{-\bar{d}}$ gives the compact set condition (3.8). By a similar calculation as in (3.20) we get that for any bounded stopping time

$$\mathrm{E}\left\|\tilde{M}_{\tau+s}^N - \tilde{M}_\tau^N\right\|^2_{\mathscr{H}^{-(\bar{d}+1)}} \leq Cs,$$

where C does not depend on s and N, which proves the Aldous criterion and thereby tightness of $\tilde{M}^N$.

To identify the limit, take $h \in \mathscr{H}^{-(\bar{d}+1)}$ and note that from the convergence of ξ_N to μ it follows that the predictable quadratic variation (3.19) converges in law to (3.17). Knowing that $(\tilde{M}^N)$ are uniformly integrable and that jumps are bounded by $\frac{1}{N}\|h\|_\infty$ we use Theorem 3.3 to obtain convergence of $\tilde{M}^N(\phi)$ to a continuous Gaussian martingale with quadratic variation given by (3.17). This identify every limit point of $(\tilde{M}^N)$ as an $\mathscr{H}^{-(\bar{d}+1)}$-valued Gaussian martingale with the quadratic covariation given by (3.17). $\square$

We are ready to prove the convergence.

Proof of Theorem 3.4 We need to check the tightness of (η_t^N) and identify the limit. By (3.21), we have

$$\mathrm{E}\sup_{0 \le t \le T} \|\eta_t^N\|^2_{\mathscr{H}^{-\bar{d}}} \le$$

$$3\mathrm{E}\|\eta_0^N\|^2_{\mathscr{H}^{-\bar{d}}} + 3\mathrm{E}\sup_{0 \le t \le T}\left\|\int_0^t (b-d)\eta_s^N \, ds\right\|^2_{\mathscr{H}^{-\bar{d}}} + 3\mathrm{E}\sup_{0 \le t \le T}\|\tilde{M}_t^N\|^2_{\mathscr{H}^{-\bar{d}}}, \tag{3.22}$$

Since $\|\eta\|_{TV} = \|\eta\|_{C_b(\mathscr{X})'}$ and by the Sobolev inequality $\|\eta\|_{\mathscr{H}^{-\bar{d}}} \le C\|\eta\|_{C_b(\mathscr{X})'}$, the first term on the right-hand side is bounded uniformly in N by assumption of the Theorem. The last term is bounded by (3.20). Note that by the Parseval identity in the base of bounded continuous functions we have

$$\|(b-d)\eta\|^2_{\mathscr{H}^{-\bar{d}}} = \sum_i \langle \phi, (b-d)\eta \rangle^2 \le (\|b\|_\infty + \|d\|_\infty)^2 \le \|\eta\|^2_{\mathscr{H}^{-\bar{d}}}.$$

So, we can estimate the middle term of (3.22) by properties of the Bochner integral and the Jensen inequality,

$$\mathrm{E}\sup_{0 \le t \le T}\left\|\int_0^t (b-d)\eta_s^N \, ds\right\|^2_{\mathscr{H}^{-\bar{d}}} \le T\mathrm{E}\sup_{0 \le t \le T}\int_0^t \|(b-d)\eta_s^N\|^2_{\mathscr{H}^{-\bar{d}}} \, ds$$

$$= T\mathrm{E}\int_0^T \|(b-d)\eta_s^N\|^2_{\mathscr{H}^{-\bar{d}}} \, ds.$$

Since $b-d$ is $C_b^{\bar{d}}$ we have $\|(b-d)h\|_{\mathscr{H}^{\bar{d}}} \le c\|h\|_{\mathscr{H}^{\bar{d}}}$ for some c and thereby

$$\|(b-d)\eta_s^N\|_{\mathscr{H}^{-\bar{d}}} \le c\|\eta_s^N\|_{\mathscr{H}^{-\bar{d}}}. \tag{3.23}$$

So

$$\mathrm{E}\sup_{0 \le t \le T}\left\|\int_0^t (b-d)\eta_s^N \, ds\right\|^2_{\mathscr{H}^{-\bar{d}}} \le cT\mathrm{E}\int_0^T \sup_{0 \le s \le T}\|\eta_s^N\|^2_{\mathscr{H}^{-\bar{d}}} \, ds$$

3.4 CLT-Type Theorems for IBM

and by Grönwall's inequality $\mathrm{E}\sup_{0\le t\le T}\left\|\int_0^t (b-d)\eta_s^N ds\right\|_{\mathscr{H}^{-\bar d}}^2$ is bounded uniformly in N. By Markov's inequality with high probability η_s^N is bounded in $\mathscr{H}^{-\bar d}$ and since $\mathscr{H}^{-\bar d}$ is compactly imbedded in $\mathscr{H}^{-(\bar d+1)}$, we have (3.8). To check the Aldous condition for (η_t^N) we use again the form (3.21). η_0^N converges by assumption, so it is tight. We have already checked tightness for the martingale part in Lemma 3.6. For the integral term, take a stopping time τ and by (3.23) we have

$$\left\|\int_0^{\tau+t}(b-d)\eta_s^N ds - \int_0^{\tau}(b-d)\eta_s^N ds\right\|_{\mathscr{H}^{-\bar d}} \le \int_{\tau}^{\tau+t}\|(b-d)\eta_s^N\|_{\mathscr{H}^{-\bar d}}ds$$

$$\le c\int_{\tau}^{\tau+t}\|\eta_s^N\|_{\mathscr{H}^{-\bar d}}ds \le ct\sup_{0\le s\le T}\left\|\eta_s^N\right\|_{\mathscr{H}^{-\bar d}}$$

which has a bounded expectation, so again by Markov inequality we have (3.10).

Now, we know that every subsequence of (η_t^N) has a convergent subsequence. Slightly abusing notation, we will denote this convergent subsequence still by (η_t^N) and show that the limit satisfies (3.16). Since $(b-d)\eta_t$ is bounded in $\mathscr{H}^{-(\bar d+1)}$, the Bochner integral in (3.16) makes sense, so it is sufficient to note that, by (3.14) for $h\in\mathscr{H}^{\bar d+1}$ we have

$$\langle h,\eta_t\rangle - \langle h,\eta_0\rangle - \int_0^t \langle h,(b-d)\eta_s\rangle ds + \langle h,\tilde M_t\rangle =$$
$$\langle h,\eta_t-\eta_t^N\rangle - \langle h,\eta_0-\eta_0^N\rangle + \int_0^t \left\langle (b-d)h,\eta_s-\eta_s^N\right\rangle ds + \langle h,\tilde M_t\rangle - \tilde M_t^{h,N}.$$

The first difference goes to zero by convergence of (η_t^N), the second by the assumption of Theorem 3.4, the third by the fact that (3.23) holds also for $\bar d+1$ and for (η_t), and the convergence of the last difference was already proved in Lemma 3.6.

The last thing to check is the uniqueness of solutions to (3.16). The way to do it is to fix a Gaussian martingale $(\tilde M_t)$ and observe that (3.16) admits pathwise unique solutions. Then we use the fact that pathwise uniqueness of solutions imply uniqueness in distribution by the Yamada-Watanabe theorem (cf. [77]).

Similarly to the classical CLT, the result of Theorem 3.4 implies the following corollary concerning the convergence rate of the IBMs (ξ_t^N) to the deterministic aproximation (ξ_t).

Theorem 3.5 *Under assumptions of Theorem 3.4 for any $\alpha > 0$ there exist $q_\alpha > 0$ and $N_0 \in \mathbb{N}$ such that for $N \ge N_0$ we have*

$$\mathrm{P}\left(\sup_{0\le t\le T}\left\|\xi_t^N-\xi_t\right\|_{\mathscr{H}^{-(\bar d+1)}} \ge \frac{q_\alpha}{\sqrt{N}}\right) \le \alpha.$$

3.5 Examples of CLTs

In this section we briefly show, what central limit-type theorems can be proved for examples of IBMs from Sect. 2.3.

Example 3.2 Let us consider now the model from Example 2.6. It differs from the previous one only by adding the intra-species competition term $\frac{1}{N}\sum_j U(x_i, x_j)$ to the death rate (cf. (2.19)). Now, the fluctuation process $\eta_t^N = \sqrt{N}\left(\xi_t^N - \xi^t\right)$ for any measurable and bounded function $h\colon \mathscr{X} \to \mathbb{R}$ satisfies

$$\langle h, \eta_t^N \rangle = \langle h, \eta_0^N \rangle + \int_0^t \langle (b-d)h, \eta_s^N \rangle ds$$
$$-\int_0^t \langle \mathscr{U}(\cdot, \xi_s^N)h, \eta_s^N \rangle ds - \int_0^t \langle \mathscr{U}(\cdot, \eta_s^N)h, \xi_s^N \rangle ds + \tilde{M}_t^{h,N},$$

where $\langle \tilde{M}^{h,N} \rangle_t = \int_0^t \langle (b-d-\mathscr{U}(\cdot, \xi_s^N))h^2, \xi_s^N \rangle ds$. So in the limit we have

$$\langle h, \eta_t \rangle = \langle h, \eta_0 \rangle + \int_0^t \langle [b-d-\mathscr{U}(\cdot, \xi_s) - \mathscr{U}(\xi_s, \cdot)]h, \eta_s \rangle ds + \tilde{M}_t^h$$

with

$$\langle \tilde{M}_t^h, \tilde{M}_t^h \rangle = \int_0^t \langle [b-d-\mathscr{U}(\cdot, \xi_s)]gh, \xi_s \rangle ds.$$

So one can expect that in an appropriate space of distributions processes η^N converge to a process satisfying

$$\eta_t = \eta_0 + \int_0^t (b-d-\mathscr{U}(\cdot, \xi_s))\eta_s \, ds + \tilde{M}_t,$$

where $\tilde{M}_t$ is a centered Gaussian martingale with $\left\langle \langle g, \tilde{M}_t \rangle, \langle h, \tilde{M}_t \rangle \right\rangle = \langle \tilde{M}_t^g, \tilde{M}_t^h \rangle$ as above. The proof can be done in the same way as the proof of Theorem 3.4 if U is sufficiently smooth.

Example 3.3 Let us consider the jump process from Example 2.4 that converges to the solution of the heat equation. As before, considering a fluctuation process $\eta_t^N = \sqrt{N}\left(\xi_t^N - \xi^t\right)$ we can write

$$\langle h, \eta_t^N \rangle = \langle h, \eta_0^N \rangle + \sqrt{N} \int_0^t \langle \sqrt{N}\left(\tfrac{1}{2}\Delta_N h - \tfrac{1}{2}\Delta h\right), \xi_s^N \rangle ds$$
$$+ \int_0^t \langle \tfrac{1}{2}\Delta h, \eta_s^N \rangle ds + \tilde{M}_t^{h,N},$$

3.5 Examples of CLTs

where $\Delta_N h(\mathbf{x}) = \sum_{k=1}^{2d} N\left(h(\mathbf{x} + \epsilon_k/\sqrt{N}) - h(\mathbf{x})\right)$ and $\tilde{M}_t^{h,N}$ is a martingale with

$$\left\langle M^{h,N}\right\rangle_t = \int_0^t \left\langle \sum_{k=1}^{2d} N\left(h(\mathbf{x} + \epsilon_k/\sqrt{N}) - h(\mathbf{x})\right)^2, \xi_s^N \right\rangle ds.$$

That means that the limit of fluctuations η_t satisfies

$$\langle h, \eta_t \rangle = \langle h, \eta_0^N \rangle + \int_0^t \left\langle \tfrac{1}{2}\Delta h, \eta_s \right\rangle ds + \tilde{M}_t^h,$$

where $\tilde{M}_t^h$ is a Gaussian martingale with

$$\left\langle M^h \right\rangle_t = \int_0^t \left\langle (\nabla h)^2, \xi_s \right\rangle ds.$$

In this case we have the Laplace operator which is not bounded on the Sobolev spaces $\mathcal{H}^n$, so the proof becomes more intricate. You need to use a sequence of imbedded spaces, to obtain appropriate estimates in the subsequent parts of the proof, see e.g. [77, 90, 113].

Example 3.4 We now show, again without the proof, the calculations that lead to the identification of the limit in the age-structured model from Example 2.7. If $\partial_N h(x) = \frac{1}{N}\left[h\left(x + \frac{1}{N}\right) - h(x)\right]$, then for any $h \in C_b^1([0, \infty))$ we have

$$\langle h, \xi_t^N \rangle = \langle h, \xi_0^N \rangle + \int_0^t \left\langle \partial_N h(\cdot) + b(\cdot)h(0) - d(\cdot)h(\cdot), \xi_s^N \right\rangle ds + M_t^{h,N},$$

where $M_t^{h,N}$ is a martingale with

$$\left\langle M^{h,N} \right\rangle_t = \int_0^t \left\langle \frac{1}{N^2}(\partial_N h(\cdot))^2 + \frac{1}{N}[b(\cdot)h(0) - d(\cdot)h(\cdot)], \xi_s^N \right\rangle ds. \quad (3.24)$$

So the fluctuation process $\eta_t^N = \sqrt{N}\left(\xi_t^N - \xi^t\right)$ satisfies

$$\langle h, \eta_t^N \rangle = \langle h, \eta_0^N \rangle + \int_0^t \left\langle \sqrt{N}\left(\partial_N h - h'\right), \xi_s^N \right\rangle ds + \int_0^t \left\langle h', \eta_s^N \right\rangle ds$$
$$+ \int_0^t \left\langle b(\cdot)h(0) - d(\cdot)h(\cdot), \eta_s^N \right\rangle ds + \sqrt{N} M_t^{h,N}$$

and we can expect the fluctuation processes to converge to η such that

$$\langle h, \eta_t \rangle = \langle h, \eta_0 \rangle + \int_0^t \left\langle h'(\cdot) + b(\cdot)h(0) - d(\cdot)h(\cdot), \eta_s^N \right\rangle ds + \tilde{M}_t^h$$

with $\langle \tilde{M}^h \rangle_t = \int_0^t \langle b(\cdot)h(0) - d(\cdot)h(\cdot), \xi_s \rangle \, ds$. Note that the coefficient $\frac{1}{N^2}$ by ∂_N in (3.24) results in vanishing the differential term in $\langle \tilde{M}_t^h \rangle$ (cf. Example 3.5).

Example 3.5 Let us now come back to the simplest example from the previous section, namely Example 2.3. The situation here is a bit different, namely even if we consider a single particle without rescaling we obtain in the limit a deterministic movement of a unit mass $\delta_{\mathbf{x}_t}$ according to the differential equation $\mathbf{x}' = \mathbf{b}(\mathbf{x})$. Therefore, when we add the rescaling by $1/N$ and increase the number of particles, we increase the rate of convergence. Note that in this case

$$\langle h, \xi_t^N \rangle = \langle h, \xi_0^N \rangle + \int_0^t \langle A_N h, \xi_s^N \rangle \, ds + M_t^{h,N},$$

where $A_N h(\mathbf{x}) = N \left(h\left(\mathbf{x} + \mathbf{b}(\mathbf{x})/N\right) - h(\mathbf{x}) \right)$ and $M_t^{h,N}$ is a martingale with

$$\langle M^{h,N} \rangle_t = \int_0^t \frac{1}{N^2} \langle (A_N h)^2, \xi_s^N \rangle \, ds.$$

Therefore,

$$\langle \sqrt{N} M_t^{h,N} \rangle = \frac{1}{N} \int_0^t \langle (A_N h)^2, \xi_s^N \rangle \, ds \to 0$$

and thereby $\eta_t^N = \sqrt{N} \left(\xi_t^N - \xi_t \right)$ vanishes in limit. We need to use different rescaling. Namely, if we define $\eta_t^N = N \left(\xi_t^N - \xi_t \right)$, then

$$\langle h, \eta_t^N \rangle = \langle h, \eta_0^N \rangle + \int_0^t \langle A_N h, \eta_s^N \rangle \, ds + \tilde{M}_t^{h,N},$$

where $\tilde{M}_t^{h,N} = N M_t^{h,N}$ and $\langle \tilde{M}^{h,N} \rangle_t = N^2 \langle M^{h,N} \rangle_t = \int_0^t \langle A_N h^2, \xi_s^N \rangle \, ds$. So we can expect that

$$\langle h, \eta_t \rangle = \langle h, \eta_0 \rangle + \int_0^t \langle \frac{1}{2} D^2 h(\mathbf{b}, \mathbf{b}), \xi_s \rangle \, ds + \int_0^t \langle \mathbf{b} \cdot \nabla h, \eta_s \rangle \, ds + \tilde{M}_t^h$$

with $\langle \tilde{M}^h \rangle_t = \int_0^t \langle \mathbf{b} \cdot \nabla h, \xi_s \rangle \, ds$.

Chapter 4
Selected Superprocesses

In this chapter we present the Dawson–Watanabe superprocess. It is the limit process for an IBM, describing cells which move according to a Brownian motion and die or divide. We introduce a stochastic partial differential equation related to this superprocess. We also present a historical superprocess and the Fleming-Viot superprocess, which combines Moran's model of genetic drift with mutations.

4.1 The Dawson–Watanabe Superprocess

In Example 2.8 we have seen that even for a relatively simple individual model a limit passage can lead to a *superprocess*, a stochastic process with values in the measure space $\mathcal{M}$ or $\mathcal{M}_1$. Due to the numerous applications of such processes, their theory has been intensively developed and is already the subject of several monographs and numerous publications [34, 38, 40, 41, 85, 101]. We now present the definition and properties of one of the basic superprocesses, the Dawson–Watanabe process, which we use in phytoplankton models.

The starting point is the process resulting from the combination of the process presented in Examples 2.4 and 2.8. Instead of a phenotypic model, we consider a population of cells moving in Brownian motion in the space $\mathcal{X} = \mathbb{R}^d$. Each cell can either divide with intensity α or die also with intensity α. Let us denote by $\mathbf{x}_t^i$ the location of i-th cell. The evolution of the population is described by a Markov process with values in the space $\mathcal{N}$ defined by the formula

$$\xi_t^1 = \sum_i \delta_{\mathbf{x}_t^i}, \quad t \geq 0. \tag{4.1}$$

© The Author(s), under exclusive license to Springer Nature Switzerland AG 2024
R. Rudnicki, R. Wieczorek, *Individual-Based Models and Their Limits*,
SpringerBriefs in Mathematical Methods,
https://doi.org/10.1007/978-3-031-75270-4_4

Let $\mathscr{P}: [0, \infty) \times \mathscr{N} \times \mathscr{B}(\mathscr{N}) \to [0, 1]$ be the transition probability function for $(\xi_t^1)_{t\geq 0}$. Then $\mathscr{P}$ satisfies (2.32). The process $(\xi_t^1)_{t\geq 0}$ is an example of a *branching Brownian motion*. Let Δ be the Laplace operator in $\mathbb{R}^d$, and $\mathfrak{D}(\Delta)$ its domain. The following theorem characterises the distribution of the process $(\xi_t^1)_{t\geq 0}$.

Theorem 4.1 *Let $v_0 \in C_b^+(\mathbb{R}^d) \cap \mathfrak{D}(\Delta)$ and $\xi_0^1 = \mu \in \mathscr{N}$. If $v(t, x)$ is a solution to the problem*

$$\frac{\partial v}{\partial t} = \frac{1}{2}\Delta v + \alpha(v-1)^2, \quad v(0, x) = v_0(x), \tag{4.2}$$

then

$$\mathrm{E}\exp\left(\langle \ln v_0, \xi_t^1 \rangle\right) = \exp\left(\langle \ln v(t, \cdot), \mu \rangle\right). \tag{4.3}$$

We now consider a sequence of processes $(\xi_t^N)_{t\geq 0}$ describing a cellular population. We assume that the cells move according to a Brownian motion and, as in Example 2.8 we assume that in the N-th process a single cell has a measure $\frac{1}{N}$, and the division and death rates are αN. The last assumption means that in the N-th process we jump from the measure $\mu = \frac{1}{N}\sum_i \delta_{x_i}$ to the measure $\mu - \frac{1}{N}\delta_{x_i}$ with intensity αN and to the measure $\mu + \frac{1}{N}\delta_{x_i}$ also with intensity αN. We then pass to the limit as $N \to \infty$. In the limit, we obtain a process $(\xi_t)_{t\geq 0}$ with values in the space $\mathscr{M}$ called the *Dawson–Watanabe superprocess* or *super-Brownian motion*.

By approximating the Brownian motion process by a random walk on a lattice, we can find the values of the generator of the Dawson–Watanabe process on F_h functions (see Examples 2.4 and 2.8) and thus determine its distributions. The distributions of the process $(\xi_t)_{t\geq 0}$ can also be determined using the following theorem.

Theorem 4.2 *Let $h \in C_b^+(\mathbb{R}^d) \cap \mathfrak{D}(\Delta)$ and $\xi_0 = \mu \in \mathscr{M}$. If $u(t, x)$ is a solution of the problem*

$$\frac{\partial u}{\partial t} = \frac{1}{2}\Delta u - \alpha u^2, \quad u(0, x) = h(x), \tag{4.4}$$

then

$$\mathrm{E}\exp\left(\langle -h, \xi_t \rangle\right) = \exp\left(\langle -u(t, \cdot), \mu \rangle\right). \tag{4.5}$$

The Dawson–Watanabe process can also be defined by a martingale problem.

Theorem 4.3 *For each function $h \in \mathfrak{D}(\Delta)$ the process*

$$M_t(h) = \langle h, \xi_t \rangle - \langle h, \xi_0 \rangle - \int_0^t \left\langle \tfrac{1}{2}\Delta h, \xi_s \right\rangle ds \tag{4.6}$$

4.1 The Dawson–Watanabe Superprocess

is a martingale with quadratic variation

$$\langle M(h)\rangle_t = 2\alpha \int_0^t \langle h^2, \xi_s\rangle \, ds. \tag{4.7}$$

The Dawson–Watanabe process can be represented as a solution of a certain stochastic partial differential equation (SPDE). In this section we limit ourselves to an introduction to this subject; a more complete presentation is in the Walsh monograph [109]. A natural way to represent the process $(\xi_t)_{t\geq 0}$ as a solution to a SPDE would be to look for a random measure $M(dx, dt)$ defined on the space $\mathbb{R}^d \times [0, \infty)$ and satisfying the condition

$$M_t(h) = \int_0^t \int_{\mathbb{R}^d} h(x)\, M(dx, ds). \tag{4.8}$$

We can assume that $M(A \times [0, t]) = M_t(\mathbf{1}_A)$ for $A \in \mathscr{B}(\mathbb{R}^d)$. The function so defined can be extended to an additive function on the σ-algebra $\mathscr{B}(\mathbb{R}^d \times [0, \infty))$. This extension is not a σ-additive function, because σ-additivity of M on sets of the form $\mathbb{R}^d \times [t_1, t_2)$ would imply a finite variation of the Feller diffusion process. A similar problem occurred in the definition of the Itô integral, which is overcome by using the finiteness of the quadratic variation of the Wiener process, and in the case of the operator M_t we replace the usual σ-additivity by introducing the notion of σ-finite measure with values in L^2 (see [41, 109]). Let $\mathscr{A} = \mathscr{B}(\mathbb{R}^d)$ and $M_t(A) = M(A \times [0, t])$ for $A \in \mathscr{A}$. With a properly selected filtration $(\mathscr{F}_t)$ the process $(M_t(A), \mathscr{F}_t, t \geq 0, A \in \mathscr{A})$ is an orthogonal L^2-martingale measure. We will not use the definitions of these terms now and will provide them in the next section. We obtain from the formulae (4.6) and (4.8) the following equation

$$\langle h, \xi_t\rangle = \langle h, \xi_0\rangle + \int_0^t \langle \tfrac{1}{2}\Delta h, \xi_s\rangle \, ds + \int_0^t \int_{\mathbb{R}^d} h(x)\, M(dx, ds). \tag{4.9}$$

We will also use the shortened notation of the Eq. (4.9):

$$\dot\xi_t = \tfrac{1}{2}\Delta \xi_t + \dot M_t. \tag{4.10}$$

We will continue to consider further the relationship of the Dawson–Watanabe process to the SPDE in Sect. 4.2.

So far we have considered a branching Brownian motion in which the population changes according to a branching process with intensity 2α and a generating function $\psi(z) = \tfrac{1}{2}(1+z^2)$. One can consider a branching Brownian motion in which the population size changes according to any branching process with continuous time. For simplicity, we consider a branching process in which any individual dying gives birth to a random number of descendants. Let a be the death rate of a single individual and (q_k) the distribution of the number of descendants and let $\psi(z) = \sum_{k=0}^\infty q_k z^k$ be the generating function of this distribution. So far we have been interested in the case: $a = 2\alpha$, $q_0 = q_2 = 1/2$ and $q_k = 0$ for the remaining k. Instead of the Brownian motion, one can consider any diffusion process [40] defined on a subsets of $\mathbb{R}^d$, or even any time-homogeneous Markov process with a generator L. With such a general definition, problems may arise with the existence of the process $(\xi_t^1)_{t \geq 0}$, because both the branching process and the diffusion process need not exist for all $t \geq 0$. We usually solve these problems by adding a death state or, in the case of domains with a boundary, by assuming that the trajectory after reaching the boundary remains at the boundary point. In the general case, the Eq. (4.2) takes the form:

$$\frac{\partial v}{\partial t} = Lv + a(\psi(v) - v). \tag{4.11}$$

We now consider a sequence of branching Brownian motions $(\xi_t^N)_{t \geq 0}$ assuming, as before, that in the N-th process a single individual has measure $\frac{1}{N}$ and the death rate is aN. The limit of the sequence $(\xi_t^N)_{t \geq 0}$ is the Dawson–Watanabe process under the assumption that the branching process under consideration is critical and has a finite variance, so $\psi'(1) = 1$ and $\sigma^2 = \psi''(1) < \infty$. In this case, the expression α in the formulae (4.4) and (4.7) are replaced by $\frac{1}{2}a\sigma^2$.

Much more general versions of the Dawson–Watanabe process are considered. These are the limits of processes $(\xi_t^N)_{t \geq 0}$, in which the population changes according to branching processes depending on N and x [40]. Let

$$\psi_{x,N}(z) = \sum_{k=0}^{\infty} q_k^{x,N} z^k, \quad \psi'_{x,N}(1) = 1 + \frac{\gamma_N(x)}{N}, \quad \psi''_{x,N}(1) = m_N(x).$$

We assume that $\gamma_N \to \gamma$ and $m_N \to m$ uniformly, the limit functions γ i m are continuous and bounded, $m(x) > 0$ and $\psi'''_{x,N}(1) \leq M$ for some $M > 0$ and any x and N. If $\beta(x) = a\gamma(x)$, $\alpha(x) = \frac{1}{2}am(x)$ and the function u is the solution of the problem

$$\frac{\partial u}{\partial t} = Lu + \beta(x)u - \alpha(x)u^2, \quad u(0, x) = h(x), \tag{4.12}$$

then the limit process $(\xi_t)_{t \geq 0}$ satisfies the problem

$$\mathrm{E}\exp\left(\langle -h, \xi_t \rangle\right) = \exp\left(\langle -u(t, \cdot), \mu \rangle\right), \quad \xi_0 = \mu \in \mathcal{M}.$$

The transition probability function $\mathscr{P} : [0, \infty) \times \mathcal{M} \times \mathscr{B}(\mathcal{M}) \to [0, 1]$ of the process $(\xi_t)_{t \geq 0}$ satisfies (2.32). Like diffusion processes, the process $(\xi_t)_{t \geq 0}$ has a version with continuous trajectories.

4.2 The Dawson–Watanabe Superprocess as a Solution of a SPDE

In the previous section we sketched how a process satisfying the martingale problem (4.6)–(4.7) can be represented as a solution of a SPDE with respect to some L^2-martingale measure closely related to this process. Our aim is to explain when the Dawson–Watanabe process can be represented by a SPDE with white noise. We will reduce the presentation of the SPDE theory to the minimum necessary. The book [31] contains an extensive overview of this subject. A brief introduction to the theory of SPDEs can be found in studies [15, 56] available on the Internet. In our considerations we use mainly the monograph by Walsh [109] and partly [41, 85].

Let $(E, \mathscr{E}, \mu)$ be a measurable space with a σ-finite measure μ and let $\mathscr{E}_b$ be a family of subsets of $\mathscr{E}$ consisting of sets of finite measure μ. Random function $W \colon \mathscr{E}_b \to \mathbb{R}$ is called *white noise* if

(1) the random variable $W(A)$ has a normal distribution $\mathscr{N}(0, \mu(A))$,
(2) if $A, B \in \mathscr{E}_b$ and $A \cap B = \emptyset$, then the random variables $W(A)$ and $W(B)$ are independent and $W(A \cup B) = W(A) + W(B)$.

A Gaussian process indexed by sets $A \in \mathscr{E}_b$ satisfying conditions (1), (2) exists because its covariance function

$$K(A, B) = \mathrm{E}(W(A) W(B)) = \mu(A \cap B)$$

is positive-definite.

Our aim is to introduce a rather general definition of stochastic integral, so that it includes, among other things, integration with respect to martingale measure associated with the Dawson–Watanabe process as well as with respect to white noise. Let $(\Omega, \mathscr{F}, \mathrm{P})$ be a probability space and $(E, \mathscr{E})$ be the standard space. In the cases we are considering $E = [0, \infty) \times \mathbb{R}^d$.

Let $U(A, \omega)$ be a real function defined on $\mathscr{A} \times \Omega$, where $\mathscr{A} \subset \mathscr{E}$ is an algebra of sets that satisfy the condition $\mathrm{E}(U^2(A)) < \infty$. We say that the map U is σ-*finite* in $L^2 = L^2(\Omega, \mathscr{F}, \mathrm{P})$ if there exists an increasing sequence of sets $E_n \in \mathscr{E}$, whose sum is E and for each n we have

$$\mathscr{E}_n \subseteq \mathscr{A}, \quad \sup\{\mathrm{E}\, U^2(A) \colon A \in \mathscr{E}_n\} < \infty,$$

where $\mathscr{E}_n = \{E_n \cap A \colon A \in \mathscr{E}\}$. Additive σ-finite function U is called *countable additive* on $\mathscr{E}_n$ (as a set function with values in L^2), if for each decreasing sequence (A_j) of elements of $\mathscr{E}_n$ fulfilling the condition $\bigcap_{j=1}^{\infty} A_j = \emptyset$ we have

$$\lim_{j \to \infty} \mathrm{E}\, U^2(A_j) = 0.$$

If for every n the function U is countably additive on $\mathscr{E}_n$, then we can extend it to the sets from $\mathscr{E}$ by taking

$$U(A) = \lim_{n \to \infty} U(A \cap E_n),$$

when this limit exists in L^2 and leaving $U(A)$ undefined otherwise. The set function extended in this way is called a σ-*finite measure with values in* L^2.

Let $(\mathscr{F}_t)$ be a right-hand continuous filtration. The process $(M_t(A), \mathscr{F}_t, t \geq 0, A \in \mathscr{A})$ is called L^2-*martingale measure* if

(a) $M_0(A) = 0$,
(b) M_t is a σ-finite measure with values in L^2 for $t > 0$,
(c) $(M_t(A))_{t>0}$ is a martingale with respect to $(\mathscr{F}_t)_{t>0}$.

The process (M_t) is called an *orthogonal martingale measure* if for any two disjoint sets A and B in $\mathscr{A}$ the process $M_t(A)M_t(B)$ is a martingale. For an orthogonal martingale measure (M_t) there exists a family of random measures $(\nu_t)_{t\geq 0}$ on $(E, \mathscr{E})$, for which the function $t \to \nu_t(A)$ is non-decreasing and right-hand continuous and $M_t^2 - \nu_t$ is a martingale with respect to the filtration $(\mathscr{F}_t)$, thus the process $t \mapsto \nu_t(A)$ is a quadratic variation of the martingale $(M_t(A))$. The assumption that the process $M_t(A)M_t(B)$ is a martingale for disjoint sets A and B is equivalent to the assumption that the skew bracket $\langle M(A), M(B) \rangle_t$ vanishes. Having a family of measures $(\nu_t)_{t \geq 0}$ we can define a measure by the formula $\nu(A \times (0, t]) = \nu_t(A)$ and extend it to $\mathscr{E} \times \mathscr{B}$, where $\mathscr{B} = \mathscr{B}((0, \infty))$.

We will now define the stochastic integral. The notion of stochastic integral can be introduced for a rather wide class of martingale measures [109]. We restrict ourselves to the definition of the integral with respect to an orthogonal martingale measure $(M_t(A), \mathscr{F}_t, t \geq 0, A \in \mathscr{A})$. We will also restrict ourselves to a finite interval $(0, T]$ and one of the sets E_n. The integral in the case of $(0, \infty)$ and E can be obtained by the limit passage $T \to \infty$ and $n \to \infty$.

Define the *covariation function* $K_t(A, B) = \langle M(A), M(B) \rangle_t$, where $\langle M(A), M(B) \rangle_t$ is the skew bracket of the martingales $(M_t(A))$ and $(M_t(B))$. Next we define a set function on rectangles by

$$K(A \times B \times (s, t]) = K_t(A, B) - K_s(A, B),$$

which can be extended to a random signed measure on $\mathscr{E} \times \mathscr{E} \times \mathscr{B}$, $\mathscr{B} = \mathscr{B}([0, \infty))$. The measure K is called the *covariation measure*. It follows from the orthogonality of the measure M that its support is concentrated on the set $\Delta(E) \times [0, \infty)$, where $\Delta(E) = \{(x, x) \colon x \in E\}$ is the diagonal in $E \times E$ and $\nu(A \times [0, t]) = K(A \times A \times [0, t])$. Then we define

$$(f, g)_K = \int_{E \times E \times \mathbb{R}_+} f(x, s) g(y, s) K(dx, dy, ds)$$

$$= \int_{E \times \mathbb{R}_+} f(x, s) g(x, s) \nu(dx, ds).$$

4.2 The Dawson–Watanabe Superprocess as a Solution of a SPDE

A function $f(x, s, \omega)$ is called *elementary* when it is of the form

$$f(x, s, \omega) = X(\omega)\mathbf{1}_{(a,b]}(s)\mathbf{1}_A(x),$$

where X is a bounded random variable $\mathscr{F}_a$ measurable and $A \in \mathscr{E}$. A finite sum of elementary functions is called a *simple function*. The class of simple functions is denoted by $\mathscr{S}$. We define on the space $E \times [0, \infty) \times \Omega$ the *predictable σ-algebra* $\mathscr{P}$ as the σ-algebra generated by $\mathscr{S}$. A function defined on $E \times [0, \infty) \times \Omega$ is called *predictable*, if it is $\mathscr{P}$-measurable. We introduce a norm in the set of predictable functions:

$$\|f\|_M = \sqrt{E(f, f)_K}.$$

We denote by $\mathscr{P}_M$ the subset of predictable functions f satisfying the inequality $\|f\|_M < \infty$. Then $(\mathscr{P}_M, \|\cdot\|_M)$ is a Banach space, and the set $\mathscr{S}$ is dense in $\mathscr{P}_M$.

The *stochastic integral* of the elementary function $f(x, s, \omega) = X(\omega)\mathbf{1}_{(a,b]}(s)\mathbf{1}_A(x)$ is defined by

$$f \bullet M_t(B) = \int_0^t \int_B f(x, s) M(dx, ds) = X(\omega)(M_{b \wedge t}(A \cap B) - M_{a \wedge t}(A \cap B)).$$

We define the integral of $f \in \mathscr{S}$ using the linearity of the integral, and then we define the integral for the function $f \in \mathscr{P}_M$ assuming that $f \bullet M_t(B) = \lim_{n \to \infty} f_n \bullet M_t(B)$, where (f_n) is any sequence of functions from the set $\mathscr{S}$ converging to f in the norm $\|\cdot\|_M$. The existence of a limit and independence of choice of the sequence (f_n) follows from the inequality

$$E\left((f \bullet M_t(B) - g \bullet M_t(B))^2\right) \le \|f - g\|_M^2 \quad \text{for } f, g \in \mathscr{S}.$$

Stochastic integral $f \bullet M_t(A)$ is an orthogonal martingale measure with covariation measure

$$K_{f \bullet M}(dx, dy, ds) = f(x, s)f(y, s) K_M(dx, dy, ds).$$

Let W be a white noise on the space $E \times [0, \infty)$ with measure μ. The process $W_t(A) = W(A \times [0, t])$ is an orthogonal martingale measure and $\nu = \mu$. The following theorem characterizes a white noise process among other orthogonal martingale measures.

Theorem 4.4 *We assume that for any set $A \in \mathscr{E}$, the function $t \to M_t(A)$ is continuous. An orthogonal martingale measure M is a white noise if and only if its covariance measure is deterministic.*

In our applications $E = \mathbb{R}^d \times [0, \infty)$, ν is the Lebesgue measure, further denoted by $|\cdot|$. Then the map $t \mapsto W_t(A)$ is an $(\mathscr{F}_t)$-Brownian motion starting at 0 with diffusion parameter $|A|$. Moreover $W_t(A \cup B) = W_t(A) + W_t(B)$ a.e. for any disjoint sets $A, B \in \mathscr{B}(\mathbb{R}^d)$ with finite Lebesgue measure and $t > 0$. We set $W_t(h) = \int_0^t \int_{\mathbb{R}^d} h(x) W(dx, ds)$ for $h \in L^2(\mathbb{R}^d)$.

We are now interested in the question of whether the Dawson–Watanabe process $(\xi_t)_{t \geq 0}$ can be represented as a solution of the SPDE with white noise. We recall that the process $(\xi_t)_{t \geq 0}$ satisfied the equation

$$\langle h, \xi_t \rangle = \langle h, \xi_0 \rangle + \int_0^t \int_{\mathbb{R}^d} h(x) M(ds, dx) + \int_0^t \langle \tfrac{1}{2} \Delta h, \xi_s \rangle ds, \qquad (4.13)$$

where the process (M_t) is an orthogonal martingale measure with quadratic variation $\langle M(h) \rangle_t = 2\alpha \int_0^t \langle h^2, \xi_s \rangle ds$. In the case $d = 1$ the measures of ξ_t have densities with respect to the Lebesgue measure, say that $\xi_t(dx) = \xi(x, t) dx$. Then the martingale integral

$$h \bullet M_t = \int_0^t \int_{\mathbb{R}^d} h(x) M(dx, ds)$$

can be written in the form

$$h \bullet M_t = \int_0^t \int_{\mathbb{R}} \sqrt{2\alpha \xi(x, s)} \, h(x) W(dx, ds), \qquad (4.14)$$

where W is a white noise in $E = \mathbb{R} \times [0, \infty)$ with Lebesgue measure. We can construct the white noise W as follows. Let $\widetilde{W}$ be a white noise independent of M. Then the white noise W is obtained using the formula

$$W_t(h) = \int_0^t \int_{\mathbb{R}} \frac{\mathbf{1}_{\{\xi(x,s) \neq 0\}} h(x)}{\sqrt{2\alpha \xi(x, s)}} M(dx, ds) + \int_0^t \int_{\mathbb{R}} \mathbf{1}_{\{\xi(x,s) = 0\}} h(x) \widetilde{W}(dx, ds),$$

so the densities of the process (ξ_t) satisfy the equation

$$\int_{\mathbb{R}} h(x) \xi(x, t) dx = \int_{\mathbb{R}} h(x) \xi(x, 0) dx + \int_0^t \int_{\mathbb{R}} \sqrt{2\alpha \xi(x, s)} \, h(x) W(dx, ds)$$

$$+ \frac{1}{2} \int_0^t \int_{\mathbb{R}} \xi(x, s) h''(x) dx \, ds.$$

Thus $\xi(x, t)$ is a weak solution of the equation

$$d\xi(x, t) = \tfrac{1}{2} \frac{\partial^2 \xi}{\partial x^2}(x, t) dt + \sqrt{2\alpha \xi(x, t)} \, dW. \qquad (4.15)$$

A more precise derivation of the Eq. (4.15) is given in the monographs [34, 85]. The equation is also studied in [31, page 232]. In the dimension $d \geq 2$ the measures ξ_t are singular with respect to the Lebesgue measure and the classical approach does not cover these cases. The corresponding SPDE for this process in higher dimensions is investigated in the work [68].

4.3 Notes on Other Superprocesses

The Dawson–Watanabe superprocess can be generalized in many ways. In Sect. 4.1 we presented a generalization in which a superprocess is obtained by a limit passage for a position-dependent branching process with arbitrary diffusion. As we have already mentioned, the diffusion process can be replaced by virtually any Markov process. For example, in [34] a family of superprocesses is studied in which we replace the Laplacian with a fractional Laplacian $-(-\Delta)^{\alpha/2}$, $0 < \alpha < 2$. Superprocesses derived from individual models with interactions are also studied. Such models will be studied in the next chapters. We also consider superprocesses in which, in addition to the independent motion of individuals, there is a stochastic motion of the medium in which the population resides. Such a superprocess can be used to describe, for example, a population of phytoplankton [101].

We mention two more superprocesses that are important from an application point of view. The first of these is a *historical superprocess*, called also a *genealogical superprocess* [34, 41]. With the approximation of the branching Brownian motion by the Dawson–Watanabe process we lose information about the genealogical structure of the population. To preserve this type of information we should know the trajectories of the individual points (individuals). For this purpose, we define *a Brownian motion with values in the space of its sample paths* by assuming, that $X_t = (W_{s \wedge t})_{s \geq 0}$, where $(W_t)_{t \geq 0}$ is a d-dimensional Wiener process. Thus, the value of a process $(X_t)_{t \geq 0}$ is the function describing its realization up to time t and constant for times $\geq t$.

The process $(X_t)_{t \geq 0}$ is a Markov process in the space $C_b([0, \infty), \mathbb{R}^d)$. The fundamental difference between the Brownian motion and the process $(X_t)_{t \geq 0}$ is that the second is not time homogeneous, which should be taken into account when defining the superprocess by means of the martingale problem. One still has to take into account the branching process describing the birth and death of individuals, and then pass to the limit as in the case of the Dawson–Watanabe process. The result is a superprocess $(H_t)_{t \geq 0}$ with values in the measure space $\mathcal{M}(C_b([0, \infty), \mathbb{R}^d))$. The Dawson–Watanabe process $(\xi_t)_{t \geq 0}$ can be regarded as the corresponding projection of the process $(H_t)_{t \geq 0}$. Let us define the mapping

$$(\pi_t \mu)(A) = \mu(f \in C_b([0, \infty), \mathbb{R}^d) : f(t) \in A), \quad A \in \mathcal{B}(\mathbb{R}^d).$$

Then $\xi_t = \pi_t H_t$.

The second superprocess, which we present based on [41], is a continuous version of Moran's genetic drift model. The model describes the evolution of the allele distribution in a population with a fixed number of individuals N in which there are a large number of genetic states. We assume that at a fixed moment, the state of an individual is a point on the lattice $\mathbb{Z}^d$. Genetic mutations correspond to a symmetric random walk on the lattice $\mathbb{Z}^d$. Furthermore, each individual is replaced by a new individual with intensity γ. The state of the new individual is drawn with the same probability $1/N$ from among the states of individuals in the population.

An individual can move in time Δt in each of $2d$ directions with probability $\Delta t/2 + o(\Delta t)$. We then rescale the movement so as to obtain a sensible limit at infinity. We assume that the step in the random walk on the lattice is of length $1/\sqrt{N}$. We assume that time runs N times faster, which is equivalent to saying that the intensity of jumps and replacing individuals is N times greater. Now the motion on the lattice approximates a Brownian motion. Let $\mathscr{L} = \mathbb{Z}^d_{1/\sqrt{N}}$ be a scaled lattice with a distance between the nearest lattice points $1/\sqrt{N}$. We assume that if $\mathbf{x}_i(t)$ is the state of i-th individual at time t, then it is represented by the measure $\frac{1}{N}\delta_{\mathbf{x}_i(t)}$. The state of the population at time t is a probabilistic measure ξ_t^N on $\mathscr{L}$:

$$\xi_t^N = \frac{1}{N}\sum_{i=1}^N \delta_{\mathbf{x}_i(t)}.$$

Then $(\xi_t^N)_{t\geq 0}$ is a jump process with the following probability of jump:

$$P(\xi_{t+\Delta t}^N = \xi_t^N - \tfrac{1}{N}\delta_{\mathbf{x}_i(t)} + \tfrac{1}{N}\delta_{\mathbf{x}_j(t)}) = \gamma\Delta t + o(\Delta t) \quad \text{for } j \neq i,$$

$$P(\xi_{t+\Delta t}^N = \xi_t^N - \tfrac{1}{N}\delta_{\mathbf{x}_i(t)} + \tfrac{1}{N}\delta_{\mathbf{x}_i(t)+\boldsymbol{\varepsilon}_k/\sqrt{N}}) = N\Delta t/2 + o(N\Delta t),$$

and the sequence $\boldsymbol{\varepsilon}_1, \ldots, \boldsymbol{\varepsilon}_{2d}$ contains all vectors of the form $\pm[0, \ldots, 0, 1, 0, \ldots, 0]$. The process (ξ_t^N) has a generator of the form

$$L_N F(\mu) = \gamma \sum_{j=1}^N \sum_{i=1}^N \left(F\left(\mu - \frac{1}{N}\delta_{\mathbf{x}_i} + \frac{1}{N}\delta_{\mathbf{x}_j}\right) - F(\mu) \right)$$

$$+ \frac{N}{2} \sum_{i=1}^N \sum_{k=1}^{2d} \left(F\left(\mu - \frac{1}{N}\delta_{\mathbf{x}_i} + \frac{1}{N}\delta_{\mathbf{x}_i+\boldsymbol{\varepsilon}_k/\sqrt{N}}\right) - F(\mu) \right).$$

If $F_h(\mu) = \exp\langle h, \mu\rangle$, then

$$L_N F_h(\mu) = F_h(\mu)\Bigg[\gamma \sum_{j=1}^N \sum_{i=1}^N \left(\exp\left(\frac{1}{N}\langle h, \delta_{\mathbf{x}_j} - \delta_{\mathbf{x}_i}\rangle\right) - 1 \right)$$

$$+ \frac{N}{2} \sum_{i=1}^N \sum_{k=1}^{2d} \left(\exp\left(\frac{1}{N}\langle h, \delta_{\mathbf{x}_i+\boldsymbol{\varepsilon}_k/\sqrt{N}} - \delta_{\mathbf{x}_i}\rangle\right) - 1 \right)\Bigg].$$

4.3 Notes on Other Superprocesses

In Example 2.4 we have checked that the second component of the sum in square brackets converges to $\frac{1}{2}\langle \Delta h, \mu \rangle$ as $N \to \infty$, where the operator Δ is a Laplacian. In order to determine the limit of the first expression, let us note that

$$\exp\left(\frac{1}{N}\langle h, \delta_{\mathbf{x}_j} - \delta_{\mathbf{x}_i}\rangle\right) - 1 = \exp\left(\frac{1}{N}(h(x_j) - h(x_i))\right) - 1$$

$$= \frac{1}{N}(h(x_j) - h(x_i)) + \frac{1}{2N^2}(h(x_j) - h(x_i))^2 + o(N^{-2}),$$

thereby

$$\sum_{j=1}^{N}\sum_{i=1}^{N}\left(\exp\left(\frac{1}{N}\langle h, \delta_{\mathbf{x}_j} - \delta_{\mathbf{x}_i}\rangle\right) - 1\right) = \frac{1}{2N^2}\sum_{j=1}^{N}\sum_{i=1}^{N}\left((h(x_j) - h(x_i))^2 + o(1)\right)$$

$$= \frac{1}{2}\iint_{\mathbb{R}^2}((h(x) - h(y))^2 \mu(dx)\mu(dy) + o(1) = \langle h^2, \mu \rangle - \langle h, \mu \rangle^2 + o(1).$$

In the limit we obtain a stochastic process $(\xi_t)_{t \geq 0}$ with values in measure space $\mathcal{M}_1(\mathbb{R}^d)$, called *Fleming–Viot superprocess* [48], whose generator L satisfies the condition

$$LF_h(\mu) = \left(\tfrac{1}{2}\langle \Delta h, \mu \rangle + \gamma \langle h^2, \mu \rangle - \gamma \langle h, \mu \rangle^2\right) F_h(h) \tag{4.16}$$

for sufficiently smooth functions h. Since we know the value of the generator on functions F_h, the formula (4.16) uniquely determines the distribution of the process $(\xi_t)_{t \geq 0}$. The values of the Dawson–Watanabe and Fleming–Viot superprocess generators on functions F_h differ only in the presence of an additional component $\langle h, \mu \rangle^2$ in (4.16).

The Fleming–Viot process can also be defined by a martingale problem.

Theorem 4.5 *For each function $h \in \mathfrak{D}(\Delta)$, the process*

$$M_t(h) = \langle h, \xi_t \rangle - \langle h, \xi_0 \rangle - \int_0^t \langle \tfrac{1}{2}\Delta h, \xi_s \rangle ds \tag{4.17}$$

is a martingale with quadratic variation

$$\langle M(h) \rangle_t = \gamma \int_0^t \left(\langle h^2, \xi_s \rangle - \langle h, \xi_s \rangle^2\right) ds. \tag{4.18}$$

Chapter 5
Phenotype Models

We present phenotypic IBMs and the limit passage from these models to a nonlinear transport equation. We study two types of models: with random or assortative mating. In the case of random mating we give a theorem on asymptotic stability of the model. We show that assortative mating can lead to a polymorphic population and sympatric speciation.

5.1 From Microscopic to Macroscopic Models

Using the paper [96] and the dissertation [113] we will now present phenotypic models that lead to the study of nonlinear stochastic semigroups. The considered models with sexual reproduction are similar to the asexual model introduced in the work of [13, 74] and studied in the paper [49]. It is worth noting that asexual models are easier to study because the phenotype is inherited from a single parent and there is no need to introduce the pairing process into the model, and the inheritance is described here by a linear operator, in contrast to a model with sexual reproduction, where a nonlinear operator appears.

We will start with an individual model describing a hermaphroditic population, so individuals have male and female reproductive organs. We assume that the phenotypic traits x are elements of some closed and convex subset $\mathscr{X}$ of $\mathbb{R}^d$ with non-empty interior. We assume that the trait of an individual does not change during its lifetime.

If a population consists of k individuals with traits $x_1, \ldots, x_k$, then its state is the measure $\nu = \delta_{x_1} + \cdots + \delta_{x_k}$. We assume that individuals i and j pair up with probability $m(x_i, x_j; \nu)\Delta t + o(\Delta t)$ in time interval Δt and produce an offspring with the trait drawn according to the distribution function $K(x_i, x_j, dy)$.

Furthermore, we introduce the natural mortality rate $D(x)$ of an individual with trait x and the competitive mortality rate $U(x, y)$ of individuals with traits x and y, so that $\sum_j U(x_i, x_j)\Delta t + o(\Delta t)$ is the probability of death of an individual with the trait x_i due to competition.

The presented description leads to a stochastic process $(\xi_t)_{t \geq 0}$ with values in the space $\mathcal{N}$ with an infinitesimal generator L defined on measurable and bounded functions $\phi \colon \mathcal{N} \to \mathbb{R}$ by the formula

$$L\phi(\nu) = \iiint_{\mathcal{X}^3} [\phi(\nu + \delta_z) - \phi(\nu)] m(x, y; \nu) K(x, y, dz) \nu(dx) \nu(dy) \qquad (5.1)$$
$$+ \int_{\mathcal{X}} [\phi(\nu - \delta_x) - \phi(\nu)] \left(D(x) + \int_{\mathcal{X}} U(x, y) \nu(dy) \right) \nu(dx).$$

The first component on the right-hand side of the formula describes the mating process and the distribution of phenotypic traits in the offspring. The process jumps from a measure ν to a measure of $\nu + \delta_z$, where z is the phenotype of the offspring. We first select with intensity $m(x, y; \nu)$ a pair of parents with phenotypes x, y, and then draw according to the distribution function $K(x, y, dz)$ the phenotype of the offspring. The second component accounts for the mortality of individuals. The process jumps here from a point (a measure) ν to the point $\nu - \delta_x$, and x is the phenotype of the dead individual.

We assume that the functions m, K, D, U satisfy the following conditions:

(A1) the function $m \colon \mathcal{X} \times \mathcal{X} \times \mathcal{M} \to [0, \infty)$ is measurable with respect to the first two variables and for any $x, y \in \mathcal{X}, \mu \in \mathcal{M}, \alpha > 0$ we have

$$m(x, y; \mu) = m(y, x; \mu), \quad m(x, y; \mathfrak{O}) = 0, \quad m(x, y; \alpha \mu) = \frac{1}{\alpha} m(x, y; \mu),$$

(A2) there is $\overline{m} > 0$, such that $m(x, y; \mu) \leq \overline{m}/\mu(\mathcal{X})$ for all $x, y \in \mathcal{X}$ and non-zero measures $\mu \in \mathcal{M}$,
(A3) for any $x, y \in \mathcal{X}$ the measure $K(x, y, \cdot)$ is probabilistic; $K(x, y, A) = K(y, x, A)$ and the function $(x, y) \mapsto K(x, y, A)$ is measurable for any set $A \in \mathcal{B}(\mathcal{X})$,
(A4) the functions $D \colon \mathcal{X} \to [0, \infty)$ and $U \colon \mathcal{X} \times \mathcal{X} \to [0, \infty)$ are continuous and bounded,
(A5) the initial distribution of ξ_0 satisfies the inequality $E\langle 1, \xi_0 \rangle < \infty$.

Theorem 5.1 *If the assumptions* (A1)–(A5) *are satisfied, then there exists a unique piecewise-constant Markov process* $(\xi_t)_{t \geq 0}$ *with values in the space* $\mathcal{N}$ *and with generator* L *defined by the formula* (5.1).

5.1 From Microscopic to Macroscopic Models

The formal proof of the Theorem 5.1 is quite long. It can be found in [113] and it is based on techniques from the work [49]. The trajectories of the process $(\xi_t)_{t\geq 0}$ can be determined using a fairly general algorithm similar to the construction of Markov chains with continuous time [42]. If the process at time $t_0 = 0$ is in state $\nu = \delta_{x_1} + \cdots + \delta_{x_k}$, we first choose a jump moment t_1 according to an exponential distribution with intensity

$$\lambda = \sum_{i=1}^{k}\sum_{j=1}^{k}\bigl(m(x_i,x_j;\nu) + U(x_i,x_j)\bigr) + \sum_{i=1}^{k} D(x_i).$$

We then draw one of two possible events: with probability $\lambda^{-1}m(x_i, x_j; \nu)$ a pair (i, j) of individuals to give birth to a child, or the death of i-th individual with probability $\lambda^{-1}\Bigl(D(x_i) + \sum_{j=1}^{k} U(x_i, x_j)\Bigr)$. In the first case, we select the phenotype z of the descendant according to the distribution $K(x, y, dz)$ and we assume that $\xi_{t_1} = \nu + \delta_z$, and $\xi_{t_1} = \nu - \delta_{x_i}$ in the second case. We follow the same procedure again starting from the measure ξ_{t_1}, etc.

Next we consider a sequence of processes $(\xi_t^N)_{t\geq 0}$, $N = 1, 2, \ldots$, constructed in the same way as the process $(\xi_t)_{t\geq 0}$ with one difference, that the mortality rate U_N due to competition is $U_N(x, y) = \frac{1}{N}U(x, y)$. Let $X_t^N = \xi_t^N/N$. Then the process $(X_t^N)_{t\geq 0}$ takes values in the space of measures $\mathscr{M}_N = \{\mu = \frac{1}{N}\nu : \nu \in \mathscr{N}\}$. We will use the following additional assumptions:

(A6) for any function $f \in C_b(\mathscr{X})$ the following mapping is continuous

$$\mathscr{X} \times \mathscr{X} \ni (x, y) \mapsto \int_{\mathscr{X}} f(z) K(x, y, dz),$$

(A7) for a fixed $\mu \in \mathscr{M}$ the mapping $\mathscr{X} \times \mathscr{X} \ni (x, y) \mapsto m(x, y; \mu)$ is continuous, and the function $m(\cdot, \cdot; \mu)$ is convergent to $m(\cdot, \cdot; \nu)$ uniformly on $\mathscr{X} \times \mathscr{X}$, when $\mu \to \nu$ weakly. Moreover, there exists a constant $L > 0$, such that for any $x, y \in \mathscr{X}$ we have $|m(x, y; \mu) - m(x, y; \nu)| \leq L\|\mu - \nu\|_{TV}$,

(A8) the sequence of initial distributions (X_0^N) is weakly convergent a.e. to some fixed measure $\mu_0 \in \mathscr{M}$,

(A9) The initial distributions (X_0^N) satisfy the inequality $\sup_{N\geq 1} \mathrm{E}\int_{\mathscr{X}} X_0^N(dx) < \infty$.

Theorem 5.2 *If the assumptions (A1)–(A9) are satisfied, then for any $T > 0$ the sequence of stochastic processes (X_t^N) converges in distribution in the space*

$\mathbb{D}([0, T], \mathcal{M})$ to a deterministic and continuous function $\mu \colon [0, T] \to \mathcal{M}$. The limit function μ is uniquely determined by the condition

$$\langle \phi, \mu_t \rangle = \langle \phi, \mu_0 \rangle + \int_0^t \iiint_{\mathscr{X}^3} \phi(z) m(x, y; \mu_s) K(x, y, dz) \mu_s(dx) \mu_s(dy) \, ds$$

$$- \int_0^t \int_{\mathscr{X}} \phi(x) \Big(D(x) + \int_{\mathscr{X}} U(x, y) \mu_s(dy) \Big) \mu_s(dx) \, ds$$

(5.2)

for any measurable and bounded function $\phi \colon \mathscr{X} \to \mathbb{R}$.

The complete proof of Theorem 5.1 is given in the dissertation [113] and it reduces to applying Corollary 8.16 of Chapter 4 of the book [42]. The idea of the proof is the following. We determine generators A and A_N, $N = 1, 2, \ldots$ of the processes (X_t) and (X_t^N), $N = 1, 2, \ldots$ Next we check that for any $F \in \mathfrak{D}(A)$ there exists a sequence (F_N) such that $F_N \in \mathfrak{D}(A_N)$, $F_N \to F$, and $A_N F_N \to AF$ with a suitable definition of the limit. Since the considered processes are solutions of martingale problems, from the above convergence it follows that the sequence of processes (X_t^N) is convergent in distribution in the space $\mathbb{D}([0, T], \mathcal{M})$.

Theorem 5.3 *We assume that the conditions* (A2), (A4) *are satisfied, the functions* m, K, D, U *are non-negative and bounded and* $K(x, y, \cdot)$ *is a probability measure. If* $\mu \colon [0, T] \to \mathcal{M}$ *is a solution to Eq.* (5.2), *then there exists* $M = M(T, \mu_0(\mathscr{X}))$ *such that* $\|\mu_t - \mu_s\|_{TV} \leq M|t - s|$ *for* $0 \leq s \leq t \leq T$.

Proof Let $0 \leq t \leq T$. From (A2) and from Eq. (5.2) we obtain

$$\mu_t(\mathscr{X}) \leq \mu_0(\mathscr{X}) + \int_0^t \iiint_{\mathscr{X}^3} m(x, y; \mu_s) K(x, y, dz) \mu_s(dx) \mu_s(dy) \, ds$$

$$\leq \mu_0(\mathscr{X}) + \overline{m} \int_0^t \iiint_{\mathscr{X}^3} \frac{K(x, y, dz)}{\mu_s(\mathscr{X})} \mu_s(dx) \mu_s(dy) \, ds$$

$$\leq \mu_0(\mathscr{X}) + \overline{m} \int_0^t \mu_s(\mathscr{X}) \, ds.$$

5.1 From Microscopic to Macroscopic Models

From the Grönwall lemma $\mu_t(\mathscr{X}) \leq e^{\overline{m}t}\mu_0(\mathscr{X})$. Let us fix a measurable function ϕ satisfying the inequality $-1 \leq \phi \leq 1$. Again using Eq. (5.2), condition (A2) and condition (A4) we obtain

$$|\langle \phi, \mu_\tau - \mu_t \rangle| \leq \int_t^\tau \iiint_{\mathscr{X}^3} m(x,y;\mu_s) K(x,y,dz)\mu_s(dx)\mu_s(dy)\,ds$$

$$+ \int_t^\tau \int_{\mathscr{X}} \left(D(x) + \int_{\mathscr{X}} U(x,y)\mu_s(dy) \right) \mu_s(dx)\,ds$$

$$\leq \int_t^\tau \left[(\overline{m}+\overline{D})\mu_s(\mathscr{X}) + \overline{U}\mu_s^2(\mathscr{X}) \right] ds.$$

Since $\mu_s(\mathscr{X}) \leq e^{\overline{m}s}\mu_0(\mathscr{X})$, we have

$$|\langle \phi, \mu_\tau - \mu_t \rangle| \leq \int_t^\tau \left[(\overline{m}+\overline{D})e^{\overline{m}s}\mu_0(\mathscr{X}) + \overline{U}e^{2\overline{m}s}\mu_0^2(\mathscr{X}) \right] ds$$

$$\leq |\tau - t| \left((\overline{m}+\overline{D})e^{\overline{m}T}\mu_0(\mathscr{X}) + \overline{U}e^{2\overline{m}T}\mu_0^2(\mathscr{X}) \right).$$

We see that $|\langle \phi, \mu_\tau - \mu_t \rangle| \leq M|\tau - t|$ for some M dependent only on T and $\mu_0(\mathscr{X})$. Inequality $\|\mu_t - \mu_s\|_{TV} \leq M|t-s|$ results from identity:

$$\|\mu_t - \mu_s\|_{TV} = \sup_{-1 \leq \phi \leq 1} |\langle \phi, \mu_t - \mu_s \rangle|.$$

Remark 5.1 It is worth noting that the process (X_t) is the limit of pure jump processes (piecewise constant), and the Eq. (5.2) contains only operators describing pure jump processes, i.e. birth of an offspring and death of an individual. In the proof that the solutions of the Eq. (5.2) satisfy the local Lipschitz condition we used only the assumptions (A2) and (A4) denoting the boundedness of the jump intensity. One should not expect similar theorems for processes in which traits change continuously, e.g., when we move from a point x to $x+t$, then $\lim_{t \to 0} \delta_{x+t} = \delta_x$ in weak convergence, but not in the norm $\|\cdot\|_{TV}$.

Instead of the integral equation (5.2), let us consider the differential equation

$$\frac{d\mu_t}{dt} = P\mu_t \tag{5.3}$$

in the space $\mathcal{M}$ with the operator $P: \mathcal{M} \to \mathcal{M}$ defined by the formula

$$P\mu(A) = \int_{\mathscr{X}} \int_{\mathscr{X}} m(x, y; \mu) K(x, y, A) \mu(dx) \mu(dy)$$
$$- \int_A \left(D(x) + \int_{\mathscr{X}} U(x, y) \mu(dy) \right) \mu(dx). \qquad (5.4)$$

If we add (A7) to the assumptions of Theorem 5.3, then the operator P satisfies the local Lipschitz condition:

$$\|P\mu - P\nu\|_{TV} \leq L(r) \|\mu - \nu\|_{TV}, \qquad (5.5)$$

when $\|\mu\|_{TV} \leq r$ and $\|\nu\|_{TV} \leq r$. Since $\|\mu_t\|_{TV} \leq e^{\overline{m}t} \|\mu_0\|_{TV}$, there exists a unique strong solution $\mu: [0, \infty) \to \mathcal{M}$ of Eq. (5.3) satisfying the initial condition $\mu_0 \in \mathcal{M}$. If we assume additionally that

(A10) $\qquad\qquad U(x, y) \geq \underline{U} > 0$ for $x, y \in \mathscr{X}$,

then $\limsup_{t \to \infty} \|\mu_t\|_{TV} \leq \overline{m}/\underline{U}$ for any solution μ_t.

5.2 CLT for a Phenotype Model

In the previous section we have seen the law of large numbers-type convergence of an individual-based phenotype-structured model to the deterministic evolution of measures described by the Eq. (5.2). As we have observed in Sect. 3.4, it is interesting in such models to study the behaviour of the fluctuation processes. In this section we present a kind of central limit theorem for the phenotype model. This will allow for some estimates on the convergence rate in Theorem 5.2.

Let

$$\eta_t^N = \sqrt{N}(\mu_t - \xi_t^N),$$

where ξ^N and μ are as in Theorem 5.2. We make the following assumptions:

(B1) the set $\mathscr{X}$ is bounded and has a Lipschitz boundary,
(B2) we assume semi-random mating, i.e. $m(x, y, \mu) = \frac{a(x,y)}{\langle a(x,\cdot),\mu\rangle}$, and there exist $0 < \underline{a} \leq \overline{a} < \infty$ such that $\underline{a} \leq a(x, y) \leq \overline{a}$ for all $x, y \in \mathscr{X}$,
(B3) there exists $\bar{K} > 0$ such that $K(x, y, \mathscr{X}) \leq \bar{K}$ for $x, y \in \mathscr{X}$,
(B4) the initial conditions satisfy:

$$\sup_{N \in \mathbb{N}} \mathrm{E}\langle 1, \xi_0^N\rangle^2 < \infty, \qquad \langle 1, \mu_0\rangle < \infty, \qquad \sup_{N \in \mathbb{N}} \mathrm{E} \|\eta_0^N\|_{\mathscr{C}'} < \infty.$$

5.2 CLT for a Phenotype Model

For two nonzero Borel measures μ, ν, let us define an operator $\mathscr{I}_{\mu,\nu}$ on bounded measurable functions $\phi: \mathscr{X} \to \mathbb{R}$ by

$$\begin{aligned}
\mathscr{I}_{\mu,\nu}\phi(w) = {}& p\int_{\mathscr{X}}\int_{\mathscr{X}} \phi(z)K(x,w,dz) \\
& \left(\frac{a(x,w)\mu(dx)}{\langle a(\cdot,w),\mu\rangle} + \frac{a(w,y)\nu(dx)}{\langle a(\cdot,y),\nu\rangle}\right) \\
& - p\int_{\mathscr{X}}\int_{\mathscr{X}}\int_{\mathscr{X}} \phi(z)K(x,y,dz)a(w,y) \\
& \frac{\mu(dx)}{\langle a(\cdot,y),\mu\rangle}\frac{\nu(dy)}{\langle a(\cdot,y),\nu\rangle} \\
& - \phi(w)\left(D(w) + \int_{\mathscr{X}} U(w,y)\nu(dy)\right) \\
& - \int_{\mathscr{X}} \phi(x)U(x,w)\mu(dx).
\end{aligned} \tag{5.6}$$

Theorem 5.4 *Let us assume (A1)–(A9) and (B1)–(B4) and let $(\mu_t)_{t\geq 0}$ be the solution to (5.2). If η_0^N converges in distribution to some η_0 in $\mathscr{H}^{-2d}$, then the sequence $(\eta^N)_{N\in\mathbb{N}}$ converges in law in $\mathbb{D}([0,T], \mathscr{H}^{-2d})$ to a $C([0,T], \mathscr{H}^{-2d})$-valued stochastic process $(\eta_t)_{t\geq 0}$, which is a unique solution to the following stochastic equation*

$$\eta_t = \eta_0 + \int_0^t \mathscr{I}_s^* \eta_s\, ds + M_t; \tag{5.7}$$

the integral is a Bochner integral, and the functional $\mathscr{I}_s^* \eta_s$ is given by

$$\mathscr{C}_0 \ni \phi \mapsto (\mathscr{I}_s^*\eta_s)(\phi) = \langle \mathscr{I}_{\mu_s,\mu_s}\phi, \eta_s\rangle \in \mathbb{R},$$

where $(Z_t)_{t\geq 0}$ is a $\mathscr{C}'$-valued centered Gaussian martingale with

$$\begin{aligned}
\langle M.(\phi), M.(\psi)\rangle_t = {}& \int_0^t \frac{p}{\langle \mu_s, 1\rangle} \int_X\int_X\int_X \phi(z)\psi(z)K(x,y,dz)\mu_s(dx)\mu_s(dy)\,ds \\
& + \int_0^t \int_X \phi(x)\psi(x)\left(D(x) + \int_X U(x,y)\mu_s(dy)\right)\mu_s(dx)\,ds
\end{aligned} \tag{5.8}$$

for $\phi, \psi \in \mathscr{C}_0$.

This result allows for the following corollary concerning the convergence rate in Theorem 5.2.

Theorem 5.5 *Under assumptions of Theorem 5.4 for any $\alpha > 0$ there exist $q_\alpha > 0$ and $N_0 \in \mathbb{N}$, such that for $N \geq N_0$ we have*

$$P\left(\sup_{0 \leq t \leq T} \|v_t^N - \mu_t\|_{\mathscr{H}^{-2d}} \geq \frac{q_\alpha}{\sqrt{N}} \right) \leq \alpha.$$

The proof of Theorem 5.4 is based on the techniques from [77] and [90] and is presented in more details in [113]. It is similar to the proof of Theorem 3.4, but requires some more technicalities.

5.3 Asymptotic Properties of a Model with Random Mating

So far we have considered a phenotypic model allowing for different forms of pairing, e.g. assortative or semi-random. We will now restrict ourselves to a model with random mating, in which an individual randomly selects a partner with intensity 1 regardless of their phenotypes. In this case $m(x, y; \mu) = 1/\mu(\mathscr{X})$. We also assume that D and U are positive constants. Then the operator P is of the form:

$$P\mu(A) = \int_{\mathscr{X}} \int_{\mathscr{X}} \frac{K(x, y, A)}{\mu(\mathscr{X})} \mu(dx)\mu(dy) - (D + U\mu(\mathscr{X}))\mu(A). \quad (5.9)$$

Let μ_t be a solution to the Eq. (5.3) and let $M(t) = \mu_t(\mathscr{X})$. Then

$$M'(t) = (1 - D - UM(t))M(t)$$

and $\lim_{t \to \infty} M(t) = 0$, when $D \geq 1$ and $\lim_{t \to \infty} M(t) = \overline{M} := (1 - D)/U$, if $D < 1$. We see that a population survives when $D < 1$, and $\overline{M}$ is its limiting size and environmental capacity.

Let $\mathscr{M}_1$ be the space of probability measures on $\mathscr{B}(\mathscr{X})$. We check that the set $\mathscr{A} = \{\overline{M}\mu : \mu \in \mathscr{M}_1\}$ is a *global attractor* for Eq. (5.3), i.e. if $\mu_0 \in \mathscr{M} \setminus \{\mathfrak{O}\}$, then $\mu_t \to \mathscr{A}$, as $t \to \infty$. It turns out that there is a stronger property. The behaviour of measures on the attractor fully describes the asymptotics of μ_t for any initial measure $\mu_0 \neq \mathfrak{O}$. If we substitute $\bar{\mu}_t = \mu_t/M(t)$, then the function $t \mapsto \bar{\mu}_t$ satisfies the equation

$$\frac{d}{dt}\bar{\mu}_t = \mathscr{P}\bar{\mu}_t - \bar{\mu}_t, \quad (5.10)$$

where $\mathscr{P}\mu(A) = \int_{\mathscr{X}} \int_{\mathscr{X}} K(x, y, A)\mu(dx)\mu(dy)$. If $\bar{\mu}_0 = \mu_0/M(0)$ and a solution $\bar{\mu}_t$ of Eq. (5.10) has a partial limit μ^*, then $\overline{M}\mu^*$ is a partial limit of the solution μ_t of Eq. (5.3) satisfying the initial condition μ_0.

5.3 Asymptotic Properties of a Model with Random Mating

We shall restrict the study of asymptotic properties of the solutions of Eq. (5.10) to the one-dimensional case, i.e., when $\mathscr{X}$ is a closed interval with a non-empty interior. Even in this case it is a non-trivial problem, since the solutions determine a nonlinear stochastic semigroup, and such a semigroup may have infinitely many significantly different stationary solutions.

The formulation of the main result concerning the asymptotics of the solutions of Eq. (5.10) will be preceded by the necessary information concerning the Wasserstein distance. We refer the reader interested in further study of this subject to the books [87, 105].

For $\alpha \geq 1$ and $q \in \mathbb{R}$ we define two sets of measures:

$$\mathscr{M}_1^\alpha = \{\mu \in \mathscr{M}_1 : \int_{\mathscr{X}} |x|^\alpha \mu(dx) < \infty\},$$

$$\mathscr{M}_1^{\alpha,q} = \{\mu \in \mathscr{M}_1^\alpha : \int_{\mathscr{X}} x \mu(dx) = q\}.$$

For any measures $\mu, \nu \in \mathscr{M}_1^1$, we define the *Wasserstein distance* by the formula

$$d(\mu, \nu) = \sup_{f \in \text{Lip}_1} \int_{\mathscr{X}} f(x) (\mu - \nu)(dx), \quad (5.11)$$

where Lip_1 is the set of all continuous functions $f \colon \mathscr{X} \to \mathbb{R}$ such that $|f(x) - f(y)| \leq |x - y|$ for $x, y \in \mathscr{X}$. The Wasserstein distance can be determined using the formula:

$$d(\mu, \nu) = \int_{\mathscr{X}} |\Phi(x)| \, dx, \quad (5.12)$$

where $\Phi(x) = (\mu - \nu)(\mathscr{X} \cap (-\infty, x])$, thus $\Phi(x)$ is the difference of the cumulative distribution functions of the measures μ and ν. The proof of (5.12) is given in [105] (also in [96]).

Convergence in the Wasserstein metric implies the weak convergence of measures. The space $\mathscr{M}_1^1$ with metric d is a Polish space [87]. The convergence of a sequence of measures (μ_n) to a measure μ in the space $\mathscr{M}_1^{1,q}$ is equivalent to the following conditions (see [107] Definition 6.7 and Theorem 6.8):

$$\mu_n \to \mu \text{ weakly, as } n \to \infty \quad \text{and} \quad \lim_{R \to \infty} \limsup_{n \to \infty} \int_{\mathscr{X}_R} |x| \mu_n(dx) = 0, \quad \text{(C)}$$

where $\mathscr{X}_R := \{x \in \mathscr{X} : |x| \geq R\}$. Let us fix $q \in \mathscr{X}$, $\alpha > 1$ and $m > 0$. Let

$$\widetilde{\mathscr{M}} = \{\mu \in \mathscr{M}_1^{1,q} : \int_{\mathscr{X}} |x|^\alpha \mu(dx) \leq m\}.$$

By Markov's inequality $\mu(\{x: |x| \leq R\}) \geq 1 - m/R^\alpha$ for $\mu \in \widetilde{\mathcal{M}}$, the set $\widetilde{\mathcal{M}}$ is tight, and hence relatively compact in the topology of weak convergence. Moreover,

$$\int_{\mathcal{X}_R} |x| \mu(dx) \leq \frac{1}{R^{\alpha-1}} \int_{\mathcal{X}_R} |x|^\alpha \mu(dx) \leq \frac{m}{R^{\alpha-1}}$$

for $\mu \in \widetilde{\mathcal{M}}$, which implies the second condition in (C). Consequently, the set $\widetilde{\mathcal{M}}$ is relatively compact in $\mathcal{M}_{1,q}$ with distance d.

We assume that there exist positive constants c_1, c_2, c_3 such that

$$\int_{\mathcal{X}} |z| K(x, y, dz) \leq c_1 + c_2|x| + c_3|y|, \tag{5.13}$$

$$\int_{\mathcal{X}} z K(x, y, dz) = \frac{x+y}{2}. \tag{5.14}$$

If the measure μ has a finite first moment, then

$$\int_{\mathcal{X}} |z| (\mathcal{P}\mu)(dz) \leq c_1 + (c_2 + c_3) \int_{\mathcal{X}} |x| \mu(dx) < \infty, \tag{5.15}$$

$$\int_{\mathcal{X}} z (\mathcal{P}\mu)(dz) = \int_{\mathcal{X}} z \mu(dz). \tag{5.16}$$

With these assumptions, any solution to the Eq. (5.10) with initial condition $\mu_0 \in \mathcal{M}_1^{1,q}$ remains in the set $\mathcal{M}_1^{1,q}$ and therefore we can restrict ourselves to study solutions in this space.

Let us denote by $\mathcal{K}$ the cumulative distribution function of the measure K:

$$\mathcal{K}(x, y, z) = K(x, y, \mathcal{X} \cap (-\infty, z]).$$

Theorem 5.6 *Fix $q \in \mathcal{X}$. Suppose that*

(i) *for all $y, z \in \mathcal{X}$ the function $\mathcal{K}(x, y, z)$ is absolutely continuous with respect to x and for each $a, b, y \in \mathcal{X}$ we have*

$$\int_F \left| \frac{\partial}{\partial x} \mathcal{K}(a, y, z) - \frac{\partial}{\partial x} \mathcal{K}(b, y, z) \right| dz < 1, \tag{5.17}$$

(ii) *there are constants $\alpha > 1$, $L < 1$, and $C \geq 0$ such that for every $\mu \in \mathcal{M}_1^{\alpha,q}$ we have*

$$\int_{\mathcal{X}} |x|^\alpha \mathcal{P}\mu(dx) \leq C + L \int_{\mathcal{X}} |x|^\alpha \mu(dx). \tag{5.18}$$

5.3 Asymptotic Properties of a Model with Random Mating

Then there exists a unique measure $\mu^* \in \mathcal{M}_1^{1,q}$ such that $\mathscr{P}\mu^* = \mu^*$ and for every initial measure $\mu_0 \in \mathcal{M}_1^{1,q}$ the solution μ_t, $t \geq 0$, of Eq. (5.10) converges to μ^* in the space $\mathcal{M}_1^{1,q}$.

The full proof of Theorem 5.6 is given in [96]. We will only present its main ideas.

Proof Denote by $\mathscr{F}_q$ the set of all cumulative distribution functions of the signed measures of the form $\mu - \nu$, where $\mu, \nu \in \mathcal{M}_1^{1,q}$. We check that condition (i) implies the inequality

$$2 \int_{\mathscr{X}} \left| \int_{\mathscr{X}} \mathscr{K}(x, y, z) \, \Phi(dx) \right| dz < \int_{\mathscr{X}} |\Phi(x)| \, dx \tag{5.19}$$

for any $y \in \mathscr{X}$ and $\Phi \in \mathscr{F}_q$, $\Phi \not\equiv 0$. Then using the formulas (5.12) and (5.19) we prove that

$$d(\mathscr{P}\mu, \mathscr{P}\nu) < d(\mu, \nu) \tag{5.20}$$

for $\mu, \nu \in \mathcal{M}_1^{1,q}$, $\mu \neq \nu$. Since the solutions of Eq. 5.10 satisfy the integral equation

$$e^t \mu_t = e^r \mu_r + \int_r^t e^s \mathscr{P}\mu_s \, ds, \tag{5.21}$$

we have

$$e^t d(\mu_t, \nu_t) \leq e^r d(\mu_r, \nu_r) + \int_r^t e^s d(\mathscr{P}\mu_s, \mathscr{P}\nu_s) \, ds$$

$$< e^r d(\mu_r, \nu_r) + \int_r^t e^s d(\mu_s, \nu_s) \, ds$$

for $0 \leq r < t \leq T$, as long as $\mu_T \neq \nu_T$. From Grönwall's inequality it follows that $d(\mu_t, \nu_t) < d(\mu_r, \nu_r)$.

The next step is to check, that for every initial measure $\mu_0 \in \mathcal{M}_1^{\alpha,q}$ its orbit $\mathscr{O}(\mu_0) = \{\mu_t : t \geq 0\}$ is a relatively compact subset of $\mathcal{M}_1^{1,q}$. In the proof we use assumption (ii) and show that there exists a constant m depending only on the integral $\int_{\mathscr{X}} |x|^\alpha \mu_0(dx)$ such that

$$\int_{\mathscr{X}} |x|^\alpha \mu_t(dx) \leq m \quad \text{for } t \geq 0. \tag{5.22}$$

This means that the trajectory of $\mathscr{O}(\mu_0)$ lies in a relatively compact subset of $\mathcal{M}_1^{1,q}$ with the Wasserstein distance. Therefore, the trajectory $\mathscr{O}(\mu_0)$ is also a relatively compact subset of $\mathcal{M}_1^{1,q}$.

Let $\{S(t)\}_{t\geq 0}$ be a family of transformations of $\mathcal{M}_1^{1,q}$ defined by $S(t)\mu_0 = \mu_t$. For $\mu \in \mathcal{M}_1^{1,q}$ we define the *ω-limit set* by

$$\omega(\mu) = \{v \colon v = \lim_{n\to\infty} \mu_{t_n} \text{ for some sequence } t_n \to \infty\}.$$

If $\mu \in \mathcal{M}_1^{\alpha,q}$, then the orbit $\mathcal{O}(\mu)$ is a relatively compact subset of $\mathcal{M}_1^{1,q}$, thus $\omega(\mu)$ is a nonempty compact set and $S(t)(\omega(\mu)) = \omega(\mu)$ for $t > 0$. Let diam (A) denote the diameter of a set A. Since diam $(\omega(\mu)) = $ diam $(S(t)\omega(\mu)) < $ diam $(\omega(\mu))$ for $t > 0$, the set $\omega(\mu)$ is a singleton. Let $\omega(\mu) = \{\mu^*\}$. Then $\lim_{t\to\infty} S(t)\mu = \mu^*$, $S(t)\mu^* = \mu^*$ for $t \geq 0$, and $\mathcal{P}\mu^* = \mu^*$. From inequality (5.20) it follows that the operator $\mathcal{P}$ has the only fixed point, hence $\lim_{t\to\infty} S(t)v = \mu^*$ for any measure $v \in \mathcal{M}_1^{\alpha,q}$. Since the set $\mathcal{M}_1^{\alpha,q}$ is dense in $\mathcal{M}_1^{1,q}$ and the maps $S(t)$ are contractions, we have $S(t)\mu_0 \to \mu^*$ for any measure $\mu_0 \in \mathcal{M}_1^{1,q}$.

We can strengthen the statement of Theorem 5.6 under an additional assumption.

Theorem 5.7 *We additionally assume that for arbitrary $x, y \in \mathcal{X}$ the measure $K(x, y, dz)$ has a continuous and bounded density $k(x, y, z)$. Then for each $q \in \mathcal{X}$ and for each initial measure $\mu_0 \in \mathcal{M}_1^{1,q}$ the solution μ_t of Eq. (5.10) can be written in the form $\mu_t = e^{-t}\mu_0 + v_t$, where v_t are absolutely continuous measures and have continuous and bounded densities $v_t(x)$. The function v_t converges uniformly to an invariant density u_*, which is also continuous and bounded function.*

We now present two examples of applications of Theorems 5.6 and 5.7, in which the offspring's trait is an additive or a multiplicative perturbation of the average parental trait.

Example 5.1 Let $\mathcal{X} = \mathbb{R}$ and the random variable Z has a positive density h and satisfies conditions $\mathrm{E}\, Z = 0$ and $\mathrm{D}^2 Z < \infty$. We assume that if x and y are the traits of the parents, the offspring have the trait

$$\frac{x+y}{2} + Z.$$

Then $k(x, y, z) = h\left(z - \frac{x+y}{2}\right)$ and $\frac{\partial}{\partial x}\mathcal{K}(x, y, z) = -\frac{1}{2}h\left(z - \frac{x+y}{2}\right)$.
Condition (i) is equivalent to the inequality

$$\int_{-\infty}^{\infty} |h(z-a) - h(z-b)|\,dz < 2$$

for $a, b \in \mathbb{R}$, which follows from the positivity of the density h. We easily check that condition (ii) holds with $\alpha = 2$. If we additionally assume that the density h is

5.3 Asymptotic Properties of a Model with Random Mating

a bounded continuous function, then Theorem 5.7 holds. The invariant density u^* satisfies the equation

$$u^*(z) = \int_{\mathbb{R}} \int_{\mathbb{R}} h\left(z - \frac{x+y}{2}\right) u^*(x) u^*(y) \, dx \, dy. \tag{5.23}$$

From (5.23) we conclude that if Y_1 and Y_2 are independent copies of random variables with density u^*, and Z is a random variable with density h independent of Y_1 and Y_2, then

$$Y \stackrel{d}{=} Z + \frac{Y_1 + Y_2}{2} \tag{5.24}$$

has the density u^*. It can be checked that if $\mu_0 \in \mathcal{M}_1^{1,q}$, then $u^*(x) = f(x - q)$, where

$$f = h_0 * h_1^{*2} * h_2^{*4} * h_3^{*8} * \ldots, \quad h_n(x) = 2^n h(2^n x) \text{ for } n \geq 0, \tag{5.25}$$

and $f * g$ denotes the convolution of f and g. From (5.24) it follows immediately that $D^2 Y = 2 D^2 Z$. For instance, if Z has the normal distribution $\mathcal{N}(0, \sigma^2)$, then Y has the distribution $\mathcal{N}(q, 2\sigma^2)$.

Example 5.2 Let $\mathcal{X} = [0, \infty)$ and let Z be a random variable with values in the interval $[0, 1]$, and has a density h such that

$$\int_0^1 x h(x) \, dx = \frac{1}{2}. \tag{5.26}$$

Suppose that if x and y are parental traits, then $(x + y)Z$ is the trait of the offspring. The functions k and $\dfrac{\partial \mathcal{K}}{\partial x}$ are of the form

$$k(x, y, z) = \frac{1}{x+y} h\left(\frac{z}{x+y}\right), \tag{5.27}$$

$$\frac{\partial}{\partial x} \mathcal{K}(x, y, z) = -h\left(\frac{z}{x+y}\right) \frac{z}{(x+y)^2} \tag{5.28}$$

for $z \in [0, x + y]$ and zero otherwise. Condition (i) of Theorem 5.6 is satisfied if the inequality

$$\int_0^\infty |h(x)x - \gamma^2 h(\gamma x) x| \, dx < 1 \tag{5.29}$$

holds for every $\gamma > 0$. For example, if the support of h contains the interval $(0, \varepsilon)$, $\varepsilon > 0$, then (5.29) is fulfilled. Condition (ii) is satisfied with $\alpha = 2$. Indeed, since $0 \le Z \le 1$ we have

$$\int_0^\infty z^2 (\mathscr{P}\mu)(dz) \le \int_0^\infty \int_0^\infty \int_0^\infty \frac{z^2}{x+y} h\left(\frac{z}{x+y}\right) dz\, \mu(dx)\mu(dy)$$

$$\le \mathrm{E}\, Z^2 \int_0^\infty \int_0^\infty (x+y)^2\, \mu(dx)\mu(dy)$$

$$\le \mathrm{E}\, Z^2 \left(2q^2 + 2\int_0^\infty y^2\, \mu(dy)\right)$$

$$\le 2q^2\, \mathrm{E}\, Z^2 + L\int_0^\infty y^2\, \mu(dy),$$

where $L = 2\,\mathrm{E}\, Z^2 < 2\,\mathrm{E}\, Z = 1$. The kernel k is not a continuous function even if the density h is continuous and we cannot directly apply Theorem 5.7 here. It can be checked that if $q > 0$, then $\mu^*(\{0\}) = 0$ and prove that the invariant measure μ^* has a density u^*, which is a continuous function on the interval $(0, \infty)$. It can be verified that $\mu_t = e^{-t}\mu_0 + v_t(x)\,dx$ and v_t converges uniformly to u_* on the sets $[\varepsilon, \infty)$, $\varepsilon > 0$. In particular, if we consider Eq. (5.10) on the space $L^1[0, \infty)$, then densities converges to u_* in L^1.

Remark 5.2 Equation (5.10) with kernel k given by (5.27) is known as the *generalised Tjon–Wu equation*. If $h = \mathbf{1}_{[0,1]}$, then (5.10) is the *Tjon–Wu equation* of the energy distribution of particles in the Boltzmann equation (see [11, 67, 103]). The asymptotic stability of the Tjon–Wu equation in L^1 space was proven by Kiełek [63]. Lasota and Traple (see [72, 73]) proved stability of the general version but in the weak convergence of measures.

5.4 Notes on Other Phenotypic Models

In the previous section, we have studied asymptotic behaviour of a phenotypic macroscopic model (5.2), but we restricted our study to the case of random pairing. Assortative mating models [36, 98, 99] appear in the literature, but these models are poorly studied. In the case of models with assortative mating computer simulations show convergence of phenotypic profiles to multimodal distributions, i.e. distributions whose densities have several maxima and the distributions concentrate around these maxima. Such results suggest the statement that assortative mating can lead to polymorphic populations and consequently to the formation of new species. A mathematical proof of this fact in the case of the model (5.2) seems to be difficult. Some partial results for the model with discrete time have been obtained in [95]. We will now present this model and briefly discuss the obtained results.

5.4 Notes on Other Phenotypic Models

We again consider a hermaphroditic population and start with an individual model. We assume that the set of phenotypic traits $\mathscr{X}$ is a closed and bounded interval with a non-empty interior. In the considered model, the population always has n individuals that, after producing offspring die off. The process of sexual reproduction depends on a *preference function* $\psi: \mathscr{X} \times \mathscr{X} \to [0, 1]$, which is a measurable function. An individual with trait x plays the role of the female and with trait y the role of the male in the expression $\psi(x, y)$. Consider a population which consists of n individuals with traits $x_1, \ldots, x_n$. We assume that individuals with traits $x_i, x_j, i \neq j$ pair up with rate

$$m(x_i, x_j) = \frac{\psi(x_i, x_j) + \psi(x_j, x_i)}{n - 1}. \tag{5.30}$$

In addition, we allow self-fertilization or vegetative reproduction with rate

$$m(x_i, x_i) = 1 - \frac{1}{2} \sum_{j \neq i} m(x_i, x_j). \tag{5.31}$$

After mating/self-fertilization an offspring is born with probability 1, it survives till the next reproduction. The assumptions made imply that there will also be n individuals in the next generation. Here we are neglecting factors related to competition and adaptation to the environment. In the case of sexual reproduction, we assume that the distribution of offspring traits of individuals with x and y traits is a random variable

$$\zeta_{x,y} = \frac{x + y}{2} + \frac{|x - y|}{2} \xi_{x,y}, \tag{5.32}$$

where $\xi_{x,y}$ is a random variable with values in the interval $[-1, 1]$ such that

$$\mathrm{E}\,\xi_{x,y} = 0 \text{ and } \mathrm{E}\,\xi_{x,y}^2 \leq c < 1, \tag{5.33}$$

where c is a constant independent of x and y. It follows from the assumptions made that the offspring's trait has a distribution in the interval between the parents' traits and its expected value is equal to the parental mean trait. By $K(x, y, dz)$ we denote the distribution of the random variable $\zeta_{x,y}$. In the case of self-fertilization the offspring has the same trait as the parent.

A limit passage with the number of individuals to infinity leads to the operator P on the space $\mathscr{M}_1$:

$$P\mu(dz) = \int_{\mathscr{X}} \int_{\mathscr{X}} \psi(x, y) K(x, y, dz)\, \mu(dx)\, \mu(dy)$$
$$+ \int_{\mathscr{X}} \int_{\mathscr{X}} \left(\tfrac{1}{2}\delta_x(dz) + \tfrac{1}{2}\delta_y(dz)\right)(1 - \psi(x, y))\, \mu(dx)\, \mu(dy). \tag{5.34}$$

In (5.34) the term $\int_{\mathscr{X}} \int_{\mathscr{X}} \psi(x,y) K(x,y,dz) \mu(dx) \mu(dy)$ denotes the distribution of newborn individuals in the process of sexual reproduction while the last term is related to vegetative reproduction.

Let $\mathscr{M}_{\delta,d}$ be the subset of $\mathscr{M}_1$ composed of convex combinations of Dirac measures δ_x supported on sets of isolated points with distance at least d, i.e.

$$\mathscr{M}_{\delta,d} = \{\nu \in \mathscr{M}: \ \nu = c_1 \delta_{x_1} + \cdots + c_n \delta_{x_n}, \ |x_i - x_j| \geq d, \text{ if } i \neq j, \ n \in \mathbb{N}\}.$$

Theorem 5.8 *Assume that $\psi(x,y) > 0$ if $|x - y| < d$. Then for any measure $\mu \in \mathscr{M}$ there exists a measure $\nu \in \mathscr{M}_{\delta,d}$ such that the sequence $(P^n \mu)$ converges weakly to ν. Moreover $\int_{\mathscr{X}} x \, \mu(dx) = \int_{\mathscr{X}} x \, \nu(dx)$.*

The proof is based on the method of moments (see e.g. Section 30 of [9]) and we skip it. The question arises whether the limiting measure is not always a single Dirac delta? Indeed, under some additional assumptions it is. Let $\psi(x,y) = \varphi(|x-y|)$, where $\varphi: [0, \infty) \to [0, \infty)$ is a non-increasing function such that $\varphi(r) > 0$ for $r \in [0, d)$ and $\varphi(r) = 0$ for $r \geq d$. From Theorem 5.8 it follows immediately that if the length of the interval $\mathscr{X}$ is less than d, then the sequence of $(P^n \mu)$ converges weakly to δ_a, where $a = \int_{\mathscr{X}} x \, \mu(dx)$. If the interval $\mathscr{X}$ is longer than d and the initial measure μ is the sum of measures $\mu_1, \ldots, \mu_k$ such that $\text{diam}(\text{supp } \mu_i) < d$ for all i and the distance of the supports of the measures μ_i and μ_j is $\geq d$ for $i \neq j$, then the sequence $(P^n \mu)$ converges weakly to the measure $\nu = c_1 \delta_{a_1} + \cdots + c_k \delta_{a_k}$, where $c_i = \mu_i(\mathscr{X})$ and $a_i = \int_{\mathscr{X}} x \, \mu_i(dx)$.

In light of the above examples, another question arises. Are there an operator P and an initial measure μ whose support is an interval, such that the sequence $(P^n \mu)$ converges to a linear combination of at least two Dirac deltas? It turns out that such an example can be constructed for appropriately chosen φ and $K(x,y,dz) = \delta_{\frac{x+y}{2}}(dz)$, but the construction is quite difficult [95].

Computer simulations show that the convergence of $P^n \mu$ to a linear combination of at least two Dirac deltas occurs for a rather large class of functions φ and distributions of the phenotype K. We will limit ourselves here to two examples, and refer the interested reader to the paper [95].

The first example in Fig. 5.1 concerns the case $K(x,y,dz) = \delta_{\frac{x+y}{2}}(dz)$ and the preference function $\varphi(r) = \mathbf{1}_{[-1,1]}(r)(r-1)^2(r+1)^2$. The φ function resembles a truncated Gaussian function. The initial distribution of the phenotype is uniform on intervals $[-a, a]$ of varying length. We see that as a increases, we obtain convergence to a linear combination of an increasing number of Dirac deltas. Figure 5.2 presents a "bifurcation graph" which shows how the position and final number of Dirac deltas in the limiting measure change as a function of the support length of the initial distribution.

Very similar computer simulation results can be obtained for other K-functions, under the assumption that the offspring trait lies between the parents' traits. This assumption plays a crucial role in the proof of Theorem 5.8, and without this assumption it is difficult to expect that the phenotypes will converge to the linear combination of Dirac deltas. Computer simulations suggest that if we omit this

5.4 Notes on Other Phenotypic Models

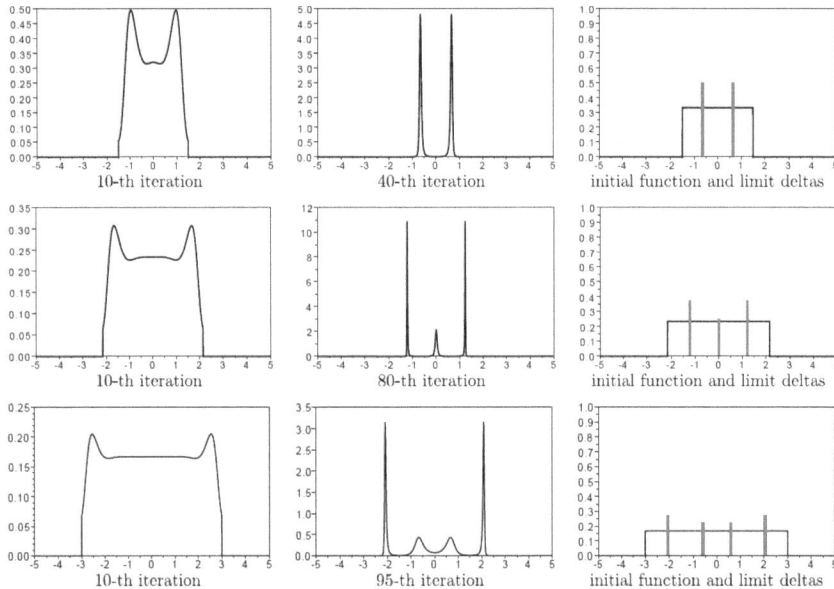

Fig. 5.1 Evolution of trait distribution with $K(x, y, dz) = \delta_{\frac{x+y}{2}}(dz)$ and $\varphi(r) = (r-1)^2(r+1)^2$. The initial trait is uniformly distributed on the intervals of length 3, 4.3, and 6 in subsequent rows

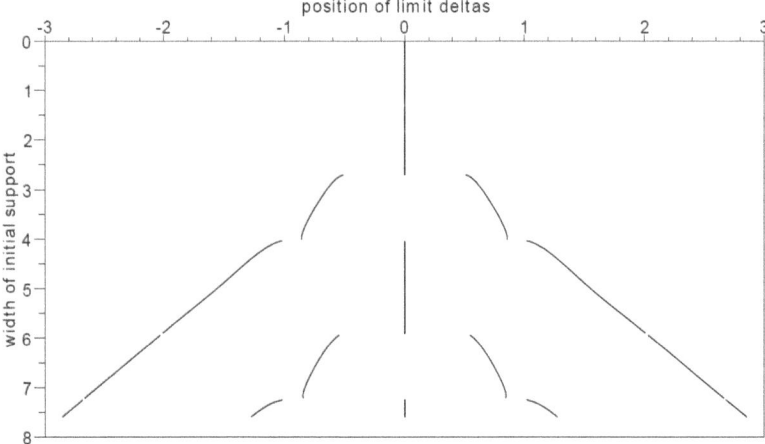

Fig. 5.2 A "bifurcation graph" of positions of the limit Dirac measures with respect to the size of the support of the initial function

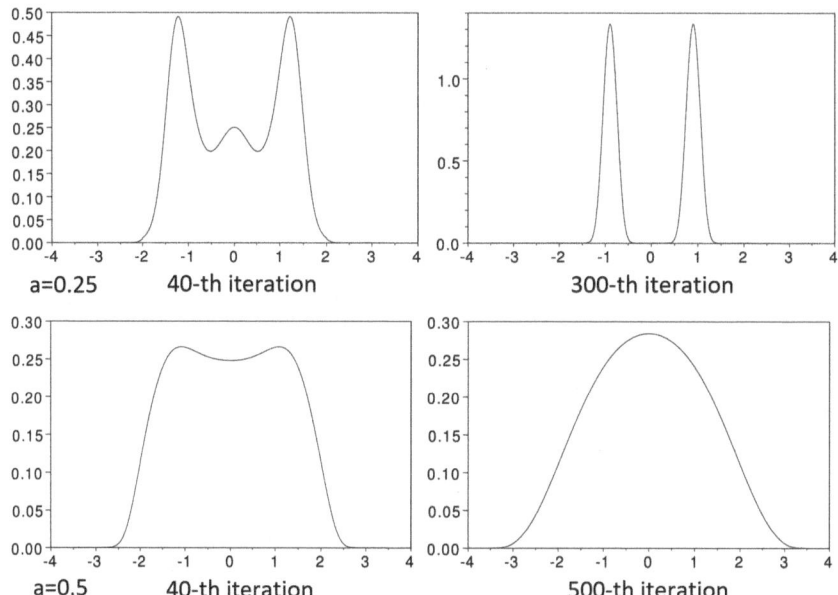

Fig. 5.3 Evolution of trait distribution with $K(x, y, dz) = \kappa \left(z - \frac{x+y}{2}\right) dz, \kappa(r) = C_a(r-a)^2(r+a)^2$ for $a = 0.25$ and $a = 0.5$. The initial distribution is uniform on the interval $[-2, 2]$

assumption, then in many cases the $(P^n \mu)$ sequences converge to a measure ν with a multimodal distribution.

Now we consider the case when the trait of the offspring after sexual mating is distributed around the average of parents' traits with some fixed symmetric distribution, that does not depend on the parental traits. Then $K(x, y, dz) = \kappa \left(z - \frac{x+y}{2}\right) dz$. Obviously, such a distribution does not satisfy the assumptions of Theorem 5.8, but seems to be quite natural [8]. We consider a model with the preceding preference function with $\kappa(r) = C_a(r - a)^2(r + a)^2$ for $|r| \le a$, where C_a is the normalizing constant. In this case, the distribution of the phenotype of the descendant lies in the interval of length $2a$. We assume that the initial distribution is a uniform distribution on the interval $[-2, 2]$. Figure 5.3 shows computer simulations for $a = 0.25$ and $a = 0.5$. For $a = 0.25$ the limiting distribution is a bimodal distribution, but already for $a = 0.5$ it is a unimodal distribution. We conclude from this that a multimodal distribution occurs when the deviation of the distribution of the offspring's trait from the parent's average is much narrower than the maximum phenotypic distance between mating individuals.

One can also consider models in which the phenotype inherited by vegetative reproduction is perturbed randomly. Here multimodal distributions may appear, but only in situations when the phenotypic trait of the offspring differs a little from that of the parent and from the average parental trait in the case of sexual reproduction.

In [112, 113] a phenotype model in bisexual populations is studied. Such a model is somewhat complicated because sex is treated as a specific phenotypic trait

5.4 Notes on Other Phenotypic Models

that significantly affects the description of mating. Under appropriate assumptions, results for hermaphroditic populations are extended to bisexual populations.

The rate of convergence of the individual (microscopic) model to the macroscopic model is investigated in [113]. This question is quite advanced mathematically. We recall that according to Theorem 5.2, the convergence of an individual model to a macroscopic model is the convergence in distribution of the sequence of processes $(\frac{1}{N}\xi_t^N)$ in the Skorohod space $\mathbb{D} = \mathbb{D}([0, T], \mathcal{M})$ to an element μ_t of this space, i.e. to some càdlàg function with values in the space of measures $\mathcal{M}$. Since stochastic processes are random variables in the space $\mathbb{D}$, Theorem 5.2 is a version of the law of large numbers in which ξ_t^N has the same role as the sum $S_N = \eta_1 + \cdots + \eta_N$ of i.i.d. random variables in the classical law of large numbers. In the case of random variables with real values, the law of large numbers means convergence $\frac{1}{N}S_N \to \mathrm{E}\,\eta_k = \frac{1}{N}\,\mathrm{E}\,S_N$, while the central limit theorem states that distributions of random variables $(S_N - \mathrm{E}\,S_N)/\sqrt{N}$ converges to a normal distribution. Since $X_t^N = \frac{1}{N}\xi_t^N$, we can expect that the sequence

$$\sqrt{N}\,(X_t^N - \mu_t)$$

is convergent to some "normal" distribution in the space $\mathbb{D}$ or in some modification of this space. This type of theorem is formulated and proved in [113].

Chapter 6
Modelling of Phytoplankton Dynamics

IBMs allow us to describe processes of formation of aggregates of phytoplankton and their movement. We present two types of phytoplankton IBMs: the first type is based on fragmentation-coagulation processes. In the second type of models, we consider single cells that move by diffusion and chemical signals. The limit passages in these models lead to nonlinear transport equations and advanced superprocesses.

6.1 Model of Growth of Phytoplankton Aggregates

Modelling the dynamics of phytoplankton growth, its movement in water, the formation of a complex spatial structure and its interaction with zooplankton is a complex issue requiring the application of various mathematical techniques [94]. Phytoplankton cells assemble into clusters called aggregates. In the process of aggregation, an important role is played by the sticky substance TEP *(Transparent Exopolymer Particles)*—a by-product of cell growth. It was also observed aggregation without the presence of TEP, which may be caused by friction due to the irregular structure of the aggregates. Aggregates are dispersed in the water column as a result of currents and turbulence, leading to a patchy distribution of phytoplankton. Phytoplankton is the basis for the vast majority of oceanic and freshwater food chains. Since the feeding organisms often have limited mobility, knowing the size and space distribution of aggregates is an important research. We present several substantially different models describing phytoplankton dynamics, which lead to non-trivial mathematical issues. The first is a microscopic model leading to a fragmentation-coagulation equation describing the size distribution of aggregates [5]. This model is a special case of the phytoplankton dynamics model considered in the work [92, 93], in which we additionally consider aggregate diffusion.

In the microscopic model, the aggregate size s is the number of phytoplankton cells of which it is composed. Denote by $\lambda_b(s)$ and $\lambda_m(s)$ the splitting rate and death rate of a single cell in an aggregate of size s, respectively. The whole aggregate can die by being submerged or devoured by animal plankton with intensity $\lambda_d(s)$. The aggregate can divide with intensity $\lambda_f(s)$ into two parts of size s_1 and s_2, where $s_1 + s_2 = s$, with probability $p(s, s_1)$. For mathematical convenience we distinguish here between the first and second parts of the aggregate, so $\sum_{s_1=1}^{s-1} p(s, s_1) = 1$. We assume that the population at time t consists of $k = k(t)$ aggregates of size $s_1, s_2, \ldots, s_k$. Any two aggregates can join to form one aggregate. The ability of an aggregate to pair with another aggregate depends on the TEP concentration, which in turn depends on the size of the aggregate s and we denote it by $c(s)$. We assume that the probability of pairing aggregates i and j per unit time is

$$\kappa_{ij} = \frac{c(s_i)c(s_j)}{\sum_{l=1}^{k} c(s_l)}. \qquad (6.1)$$

The coagulation process considered in our model differs significantly from the standard Smoluchowski coagulation model [108], where the coagulation probability is proportional to the square of the number of particles. In our case, a single aggregate of size s can combine with another with an intensity proportional to $c(s)$.

The description of phytoplankton dynamics presented above leads to an individual model, in which $\mathscr{X} = \mathbb{N}_+$. The process is a Markov chain with continuous time $(\xi_t)_{t \geq 0}$ defined on the space $\mathscr{N}$. Taking into account all possible jumps of the process $(\xi_t)_{t \geq 0}$ we easily determine its generator

$$L_1 f(\mu) = \sum_{i=1}^{k} \Big[s_i \lambda_b(s_i) f(\mu - \delta_{s_i} + \delta_{s_i+1}) + s_i \lambda_m(s_i) f(\mu - \delta_{s_i} + \delta_{s_i-1})$$

$$+ \lambda_d(s_i) f(\mu - \delta_{s_i}) + \lambda_f(s_i) \sum_{\bar{s}=1}^{s_i-1} p(s_i, \bar{s}) f(\mu - \delta_{s_i} + \delta_{\bar{s}} + \delta_{s_i-\bar{s}}) \Big]$$

$$+ \sum_{i,j=1}^{k} \frac{c(s_i)c(s_j)}{\sum_{l=1}^{k-1} c(s_l)} f(\mu - \delta_{s_i} - \delta_{s_j} + \delta_{s_i+s_j}) - \bar{\lambda}(\mu) f(\mu), \qquad (6.2)$$

where

$$\bar{\lambda}(\mu) = \sum_{i=1}^{k} \Big[s_i \lambda_b(s_i) + s_i \lambda_m(s_i) + \lambda_d(s_i) + \lambda_f(s_i) \Big] + \sum_{i,j=1}^{k} \frac{c(s_i)c(s_j)}{\sum_{l=1}^{k-1} c(s_l)},$$

$\mu = \delta_{s_1} + \cdots + \delta_{s_k}$, and f is a bounded measurable function defined on $\mathscr{N}$ with values in $\mathbb{R}$. We also assume that $L_1 f(\mathfrak{O}) = 0$, hence the zero-measure $\mathfrak{O}$ is an

6.1 Model of Growth of Phytoplankton Aggregates

absorbing state of the process (ξ_t). We assume that coefficients $\lambda_b, \lambda_m, \lambda_d, \lambda_f$ are non-negative functions and c is a positive function. For the process (ξ_t) to be non-explosive, it is sufficient to assume that the coefficient of λ_m is bounded. With this assumption, the total number of cells in the entire population is bounded a.s. at any time t. Since we will further consider models in which the aggregate sizes are positive real numbers, we assume that all coefficients are continuous, non-negative and bounded functions defined on $[0, \infty)$, and the function c is positive.

We fix a positive integer N and consider a model in which a single phytoplankton cell has size $1/N$. If $y = s/N$ is the aggregate size, then we assume that $\lambda_b(y)$, $\lambda_m(y), \lambda_d(y), \lambda_f(y), c(y)$ are the coefficients of division and death of a single cell in the aggregate, the death rate of the whole aggregate, the division rate of the aggregate and the ability of an aggregate to combine with another aggregate. An aggregate may divide into two parts of size s_1/N and s_2/N, where $s_1 + s_2 = s$, with probability $p_N(s/N, s_1/N)$. We assume that $\sum_{s_1=1}^{s-1} p_N(s/N, s_1/N) = 1$. If we have k aggregates, we assume that two aggregates of size s_i/N and s_j/N can combine into one with probability κ_{ij} given by the formula (6.1), in which s_i and s_j are replaced by s_i/N and s_j/N.

A population composed of k aggregates of size $s_1/N, \ldots, s_k/N$ is described by a measure

$$\mu = \frac{1}{N} \left(\delta_{s_1/N} + \cdots + \delta_{s_k/N} \right), \quad k \in \mathbb{N}, \ s_i \in \mathbb{N}_+ \text{ for } i = 1, \ldots, k. \quad (6.3)$$

The stochastic process presented above is a Markov chain $(\xi_t^N)_{t \geq 0}$ on the space $\mathcal{N}_N$ of measures of the form (6.3) with the generator L_N, which we obtain from the generator L_1 by replacing δ_s by $\frac{1}{N}\delta_{s/N}$, replacing the coefficients $s\lambda_b(s)$ and $s\lambda_m(s)$ by $s\lambda_b(s/N)$ and $s\lambda_m(s/N)$, and replacing s by s/N in the coefficients b_d, b_f and c.

We will now determine the limit of a sequence of processes $(\xi_t^N)_{t \geq 0}$ as $N \to \infty$. We assume that if $s_N/N \to y$ and $\bar{s}_N/N \to \bar{y}$, then

$$N p_N(s_N/N, \bar{s}_N/N) \to q(y, \bar{y})$$

and this convergence is uniform. About the function $q(y, \bar{y})$ we assume that it is nonnegative, continuous and defined for $y > 0$ and $\bar{y} \in (0, y)$ and satisfies the conditions:

$$q(y, y - \bar{y}) = q(y, \bar{y}) \quad \text{for } \bar{y} \in (0, y) \text{ and } \int_0^y q(y, \bar{y}) \, d\bar{y} = 1 \quad \text{for } y > 0.$$

From the above equations we obtain

$$\int_0^y \bar{y} q(y, \bar{y}) \, d\bar{y} = \frac{y}{2}.$$

Theorem 6.1 *We assume that the sequence (ξ_0^N) is weakly convergent to $\mu_0 \in \mathcal{M}$. Then the distributions of processes $(\xi_t^N)_{t\geq 0}$ converge weakly to a (non-random) function $t \mapsto \mu_t$ with values in the space $\mathcal{M}$ and given by the following equation*

$$\langle h, \mu_t \rangle - \langle h, \mu_0 \rangle = \int_0^t [\langle Lh, \mu_\tau \rangle + C(h, \mu_\tau)] \, d\tau \tag{6.4}$$

for $h \in C_b^1[0, \infty)$, where

$$Lh(y) = y\left(\lambda_b(y) - \lambda_m(y)\right) h'(y) - \lambda_d(y) h(y)$$

$$+ \lambda_f(y) \left(2 \int_0^y h(\overline{y}) q(y, \overline{y}) d\overline{y} - h(y) \right), \tag{6.5}$$

$$C(h, \mu) = \int_0^\infty \int_0^\infty \frac{c(y) c(\overline{y})}{\int_0^\infty c(r) \mu(dr)} [h(y + \overline{y}) - h(y) - h(\overline{y})] \mu(dy) \mu(d\overline{y}).$$

Theorem 6.1 is a special case of Theorem 1 of the paper [92], where we additionally consider the movement of phytoplankton aggregates.

If the measure μ_0 is absolutely continuous with respect to the Lebesgue measure, then also the measure μ_t is absolutely continuous and from Eq. (6.4) it follows that the function $u(t, y) = d\mu_t/dy$ satisfies the equation

$$\frac{\partial u(t, y)}{\partial t} = L^* u(t, y) + \overline{C} u(t, y), \tag{6.6}$$

where

$$L^* f(y) = \frac{d}{dy} [y (\lambda_m(y) - \lambda_b(y)) f(y)] - \lambda_d(y) f(y)$$

$$+ 2 \int_y^\infty \lambda_f(\overline{y}) f(\overline{y}) q(\overline{y}, y) \, d\overline{y} - \lambda_f(y) f(y),$$

$$\overline{C} f(y) = \int_0^y \frac{c(y - \overline{y}) c(\overline{y})}{\int_0^\infty c(r) f(r) \, dr} f(y - \overline{y}) f(\overline{y}) \, d\overline{y} - 2 c(y) f(y).$$

The properties of solutions of Eq. (6.6) are studied in the paper [5]. If $\lambda_b(y) = \lambda_m(y) + \lambda_d(y)$, then the total number of phytoplankton cells remains constant. This is an obvious conclusion from the balance of the increase in cells when they divide and the loss due to their death or the death of whole aggregates.

This property also follows from the fact that if $h(x) = x$, then $Lh \equiv 0$ and $C(h, \mu) = 0$. In this case Eq. (6.6) generates a nonlinear stochastic semigroup $\{P(t)\}_{t\geq 0}$ on the space $E = L^1(\mathbb{R}_+, \mathscr{B}(\mathbb{R}_+), y\,dy)$, which means that if $u(0, y)$ is a probability density in E, then the function $u(t, y)$ is also a density in the space E for any $t > 0$. It is often assumed that q is of the form

$$q(y, \bar{y}) = \frac{1}{y} \psi\left(\frac{\bar{y}}{y}\right) \tag{6.7}$$

for some function $\psi \colon [0, 1] \to [0, \infty)$. The assumption (6.7) is quite natural. It means that the distribution of the quotient $\bar{y}/y$ does not depend on y. If, in addition, the coefficients $\lambda_b, \lambda_f, \lambda_d, \lambda_m$ are constants and $c(y) = cy$, then we can determine the equations for the moments of the solutions and study their asymptotics [5].

6.2 CLT for Model of Growth of Phytoplankton Aggregates

Let us consider the sequence of processes $(\xi_t^N)_{t\geq 0}$ from the previous section and take

$$\eta_t^N = \sqrt{N}(\xi_t^N - \mu_t),$$

where μ_t is given by (6.4). For any measurable and bounded function $h \colon \mathbb{R}_+ \to \mathbb{R}$ we have

$$\langle h, \xi_t^N \rangle = \langle h, \xi_0^N \rangle + \int_0^t \langle L^N h, \xi_s^N \rangle \, ds + \int_0^t C^N(h, \xi_s^N) \, ds + M_t^{h,N}, \tag{6.8}$$

where

$$L^N h(x) = \lambda_b(x) N x \left(h(x + \tfrac{1}{N}) - h(x) \right) + \lambda_m(x) N x \left(h(x - \tfrac{1}{N}) - h(x) \right)$$

$$- \lambda_d(x) h(x) + \lambda_f(x) \left(2 \sum_{s=1}^{Nx-1} p_N(x, s/N) h(s/N) - h(x) \right),$$

$$C^N(h, \mu) = \sum_{i=1}^{k} \sum_{j=1}^{k} \frac{c(s_i/N)c(s_j/N)}{\sum_{l=1}^{k} c(s_l/N)} \frac{1}{N} \left[h\left(\tfrac{s_i + s_j}{N}\right) - h(s_i/N) - h(s_j/N) \right]$$

$$= \iint_{\mathbb{R}_+^2} \frac{c(x)c(y)}{\int_{\mathbb{R}_+} c(z)\mu(dz)} [h(x+y) - h(x) - h(y)] \mu(dx)\mu(dy)$$

for μ given by (6.3), and in a similar way as in Lemma 3.3 we check that $M_t^{h,N}$ is a martingale with the predictable quadratic variation

$$\left\langle M^{h,N}\right\rangle_t = \int_0^t \int_{\mathbb{R}_+} \frac{x}{N^2}\left(\lambda_b(x)(\partial_N^+ h(x))^2 + \lambda_m(x)(\partial_N^- h(x))^2\right) \xi_s^N(dx)\,ds$$

$$+ \int_0^t \int_{\mathbb{R}_+} \frac{1}{N}\lambda_d(x)h^2(x)\,\xi_s^N(dx)\,ds$$

$$+ \int_0^t \int_{\mathbb{R}_+} \frac{1}{N}\lambda_f(x) \sum_{s=1}^{Nx-1} p_N(x, \tfrac{s}{N})$$

$$\left[h(s/N) + h(x - \tfrac{s}{N}) - h(x)\right]^2 \xi_s^N(dx)\,ds$$

$$+ \iint_{\mathbb{R}_+^2} \frac{c(x)c(y)}{\int_{\mathbb{R}_+} c(z)\xi_s^N(dz)}[h(x+y)-h(x)-h(y)]^2 \xi_s^N(dx)\xi_s^N(dy)\,ds,$$

where $\partial_N^\pm h(x) = N[h(x \pm \tfrac{1}{N}) - h(x)]$. Now, subtracting (6.4) from (6.8), we get

$$\langle h, \eta_t^N \rangle = \langle h, \eta_0^N \rangle + M_t^{h,N} + \int_0^t \int_{\mathbb{R}_+} (\lambda_b(x) - \lambda_m(x))x\,h'(x)\eta_s^N(dx)\,ds$$

$$+ \int_0^t \int_{\mathbb{R}_+} \left(\lambda_b(x)x\sqrt{N}(\partial_N^+ h(x) - h'(x))\right.$$

$$\left. + \lambda_m(x)x\sqrt{N}(\partial_N^- h(x) + h'(x))\right) \xi_s^N(dx)\,ds$$

$$- \int_0^t \langle \lambda_d h, \eta_s^N \rangle\,ds + \int_0^t \int_{\mathbb{R}_+} \lambda_f(x)$$

$$\left[2\int_0^x q(x,y)h(y)\,dy - h(x)\right] \eta_s^N(dx)\,ds$$

$$+ \int_0^t \int_{\mathbb{R}_+} 2\lambda_f(x)\sqrt{N}$$

$$\left[\sum_{s=1}^{Nx} p^N(x, s/N)h(s/N) - \int_0^x q(x,y)h(y)\,dy\right] \xi_s^N(dx)\,ds$$

$$+ \int_0^t \int_{\mathbb{R}_+^2} \frac{c(x)c(y)}{\int_{\mathbb{R}_+} c(z)\xi_s^N(dz)}[h(x+y) - h(x) - h(y)]$$

$$\left(\xi_s^N(dx) + \mu_s(dx)\right) \eta_s(dy)\,ds$$

6.2 CLT for Model of Growth of Phytoplankton Aggregates

$$+ \int_0^t \int_{\mathbb{R}_+} c(z) \eta_s(dz)$$

$$\int_{\mathbb{R}_+^2} \frac{c(x)c(y)[h(x+y) - h(x) - h(y)]}{\int_{\mathbb{R}_+} c(z) \xi_s^N(dz) \int_{\mathbb{R}_+} c(z) \mu_s(dz)} \xi_s^N(dx) \mu_s(dy) \, ds.$$

The lines containing $\sqrt{N}$ tend to zero as N goes to infinity, so in the limit we get

$$\langle h, \eta_t \rangle = \langle h, \eta_0 \rangle + \int_0^t \left[\langle Lh, \eta_s \rangle + \tilde{C}(h, \mu_s, \eta_s) \right] ds + M_s^h, \qquad (6.9)$$

where the martingale (M_s^h) is the limit of $(M_s^{N,h})$, the operator L is given by (6.5) and

$$\tilde{C}(h, \mu, \eta) = 2 \iint_{\mathbb{R}_+^2} \frac{c(x)c(y)}{\int_{\mathbb{R}_+} c(z) \mu_s(dz)} [h(x+y) - h(x) - h(y)] \mu(dx) \eta(dy)$$

$$+ \int_{\mathbb{R}_+} c(z) \eta(dz)$$

$$\iint_{\mathbb{R}_+^2} \frac{c(x)c(y)[h(x+y) - h(x) - h(y)]}{\left(\int_{\mathbb{R}_+} c(z) \mu(dz)\right)^2} \mu(dx) \mu(dy).$$

If we assume some smoothness and boundedness assumptions for the coefficients, and introduce an appropriate weighted Sobolev space $\mathscr{H}$ and an appropriate subspace of continuous functions $\mathscr{C} \subset \mathscr{H}$, then, using similar methods to Theorem 5.4 and techniques from [77], we can prove the following theorem.

Theorem 6.2 *Let $(\mu_t)_{t \geq 0}$ be the solution of (6.4). If η_0^N converges in distribution to some η_0 in $\mathscr{H}'$, then the sequence $(\eta^N)_{N \in \mathbb{N}}$ converges in law in $\mathbb{D}([0, T], \mathscr{H}')$ to a $C([0, T], \mathscr{H}')$-valued stochastic process $(\eta_t)_{t \geq 0}$, which is a unique in $C([0, T], \mathscr{C}')$ solution to the following stochastic equation*

$$\eta_t = \eta_0 + \int_0^t \mathscr{J}_s^* \eta_s \, ds + M_t; \qquad (6.10)$$

the integral is a Bochner integral, and the functional $\mathscr{J}_s^* \eta_s$ is given by

$$\mathscr{C} \ni h \mapsto (\mathscr{J}_s^* \eta_s)(h) = \langle Lh, \eta_s \rangle + \tilde{C}(h, \mu_s, \eta_s) \in \mathbb{R},$$

where $(M_t)_{t\geq 0}$ is a $\mathscr{C}'$-valued centered Gaussian martingale with

$$\langle M(h)\rangle_t = \int_0^t \int_{\mathbb{R}_+} \lambda_d(x) h^2(x)\, \mu_s(dx)\, ds$$
$$+ \int_0^t \int_{\mathbb{R}_+} \lambda_f(x) \int_0^x q(x,y)[h(y) + h(x-y) - h(x)]^2\, \mu_s(dx)\, ds$$
$$+ \iint_{\mathbb{R}_+^2} \frac{c(x)c(y)}{\int_{\mathbb{R}_+} c(z)\xi_s^N(dz)}$$
$$[h(x+y) - h(x) - h(y)]^2 \xi_s^N(dx)\xi_s^N(dy)\, ds,$$

for $h \in \mathscr{C}$.

6.3 Models of Growth and Movement of Aggregates

In [92, 93] models of aggregate growth and movement due to diffusion are investigated. We assume that an aggregate of size s moves in the space $\mathbb{R}^d$ by Brownian motion with diffusion coefficient $D(s)$. If $d = 3$ and the aggregate is spherical, then the diffusion coefficient can be determined by the *Einstein formula*

$$D = \frac{k_B T}{m\beta} = \frac{k_B T}{6\pi \eta r}, \qquad (6.11)$$

where k_B is the Boltzmann constant, T is the temperature, β coefficient of friction, η is the viscosity of the water and r is the radius of the particle. We see that the coefficient $D(s)$ is proportional to $s^{-1/3}$.

As with the study of aggregate growth, the starting point is an individual model. Such a model can be constructed in two ways. The first way we have already learned in Sect. 1.2. In this case the aggregate of size s moves according to the equation $d\mathbf{x}_t = \sqrt{2D(s)}\, dW_t$ independently of the other aggregates until the jump caused by one of the five events considered in the aggregate growth model (division or death of a single cell, death of an aggregate, splitting of an aggregate, or fusion of two aggregates).

The second way is that we approximate the process of Brownian motion by a jump process. We assume an aggregate skips by a vector of the length ε in one of $2d$ directions (parallel to one of the axes) with intensity $(1/\varepsilon^2)D(s)$. In this case we obtain an individual model, in which $\mathscr{X} = \mathbb{R}^d \times \mathbb{N}_+$, and $(\xi_t)_{t\geq 0}$ is a continuous time jump process defined on the space $\mathscr{N}$ of measures of the form

$$\mu = \delta_{\mathbf{x}_1,s_1} + \cdots + \delta_{\mathbf{x}_k,s_k},$$

where $k \in \mathbb{N}$, $\mathbf{x}_i \in \mathbb{R}^d$ and $s_i \in \mathbb{N}_+$ for $i = 1,\ldots,k$.

6.3 Models of Growth and Movement of Aggregates

The presentation of both individual models needs to be completed by a description of the aggregate joining. First, the probability of aggregation depends on the distance between aggregates and their mass. We assume that the i-th and j-th aggregates join up with probability

$$\kappa_{ij} = \frac{c(s_i)c(s_j)}{\sum_{l=1}^{k} c(s_l)} v(\|\mathbf{x}_i - \mathbf{x}_j\|) \tag{6.12}$$

for some continuous and decreasing function $v \colon [0, \infty) \to [0, \infty)$.

The next problem is the position of new aggregates after fragmentation and coagulation. Let $\widehat{\mathbf{x}} = \widehat{\mathbf{x}}(\mathbf{x}_i, \mathbf{x}_j, s_i, s_j)$ be the location of an aggregate resulting from the fusion of aggregates with parameters $(\mathbf{x}_i, s_i)$ and $(\mathbf{x}_j, s_j)$. We may assume that

$$\widehat{\mathbf{x}} = \frac{\mathbf{x}_i + \mathbf{x}_j}{2}, \tag{6.13}$$

but the better physical choice is the centre of mass of both aggregates

$$\widehat{\mathbf{x}} = \frac{s_1 \mathbf{x}_1 + s_2 \mathbf{x}_2}{s_1 + s_2}. \tag{6.14}$$

We assume that if an aggregate splits, then immediately after splitting the descendant aggregates have the identical position as the aggregate before the division. Of course, one can also consider a model in which, after splitting, the position of the descendant aggregates is chosen in a different way (see [93]).

Consider a model in which we have replaced diffusion by a jump process, and the position of the aggregate after joining two others is given by the formula (6.13). Then we can determine the generator L_1 of the process. It can be obtained by using the formula (6.2). First, in the formula (6.2) the measure δ_s is replaced by $\delta_{\mathbf{x},s}$. The generator should then be completed with an expression corresponding to diffusion:

$$L^{\text{diff}} f(\mu) = \sum_{i=1}^{k} \left[\frac{D(s_i)}{\varepsilon^2} \sum_{l=1}^{2d} [f(\mu - \delta_{\mathbf{x}_i, s_i} + \delta_{\mathbf{x}_i + \boldsymbol{\varepsilon}_l, s_i}) - f(\mu)] \right],$$

and the sequence $\boldsymbol{\varepsilon}_1, \ldots, \boldsymbol{\varepsilon}_{2d}$ contains all vectors of the form $\pm \varepsilon [0, \ldots, 0, 1, 0, \ldots, 0]$. Also, the expression corresponding to coagulation should be replaced by a new one:

$$\sum_{i,j=1}^{k} \frac{c(s_i)c(s_j)v(\mathbf{x}_i - \mathbf{x}_j)}{\sum_l c(s_l)} [f(\mu - \delta_{\mathbf{x}_i, s_i} - \delta_{\mathbf{x}_j, s_j} + \delta_{(\mathbf{x}_i + \mathbf{x}_j)/2, s_i + s_j}) - f(\mu)].$$

We define the process $(\xi_t^N)_{t \geq 0}$ on the space $\mathcal{N}_N$ similarly to the aggregate growth model without diffusion with a corresponding change in the expression responsible for coagulation. In the definition of the generator of this process, we add the expression responsible for diffusion $L^{\text{diff}} f(\mu)$ assuming that in N-th process $\varepsilon = 1/N$ and before the expressions $\delta_{\mathbf{x}_i, s_i}$ and $\delta_{\mathbf{x}_i + \boldsymbol{\varepsilon}_l, s_i}$ there is the coefficient N^{-1}. The distributions of processes $(\xi_t^N)_{t \geq 0}$ are weakly convergent to a (non-random) function $t \mapsto \mu_t$ with values in the space $\mathcal{M}$. If the measure μ_0 is absolutely continuous with respect to the Lebesgue measure on $\mathbb{R}^d \times \mathbb{R}_+$, then the measure μ_t is also absolutely continuous and their densities satisfy a diffusion-growth-fragmentation-coagulation equation. Solutions of this equation define a semigroup of nonlinear operators on the space $L^1_+(\mathbb{R}^d \times \mathbb{R}_+)$ describing the evolution of the position and size distributions of aggregates [92].

In [93] some numerical simulations of individual models are presented, which show that growth, fragmentation and coagulation processes can lead to clusters of relatively large aggregates. The study of the type of spatial distribution of aggregates is carried out on the basis of *Clark–Evans index* [28], which compares the observed mean smallest distances between individuals with a random distribution.

6.4 Phytoplankton Dynamics as a Superprocess

Using the article by R. Adler [2], we now show how a simple biological description leads to a rather sophisticated mathematical model of phytoplankton dynamics. Consider a population of phytoplankton cells living in the space $\mathbb{R}^d$. We assume that at the initial time $t = 0$ we have N cells. We also assume that each cell moves according to a d-dimensional Wiener process and the movements of individual cells are independent. At successive moments in time $1/N, 2/N, 3/N, \ldots$ each cell either divides or dies and the probabilities of both events are $1/2$. The new population of cells moves according to the previous rules. Note that the process describing the population size is a critical branching process, and so the population will die out in finite time with probability one. Let $k(t)$ be the number of cells at time t, and $x_1(t), \ldots, x_{k(t)}(t)$ their positions in space. We assume that

$$X_t^N = \frac{1}{N} \sum_{i=1}^{k(t)} \delta_{x_i(t)}.$$

Then, the process $(X_t^N)_{t \geq 0}$ is weakly convergent to a process $(X_t)_{t \geq 0}$ with values in the measure space $\mathcal{M}$. The process $(X_t)_{t \geq 0}$ is the Dawson–Watanabe process studied in Sects. 4.1 and 4.2.

The resulting superprocess (X_t) has a number of interesting properties, which depend on the dimension of the space d. If $d = 1$, then the values of the process X_t are measures with densities $f(t, x)$ satisfying the stochastic partial differential equation

$$df(x, t) = \frac{1}{2} \frac{\partial^2 f}{\partial x^2}(x, t) \, dt + \sqrt{f(x, t)} \, dW, \qquad (6.15)$$

6.4 Phytoplankton Dynamics as a Superprocess

where $W(t, x)$ is a time-space Gaussian white noise. If $d > 1$, then the values of X_t are singular measures, and we replace Eq. (6.15) with its martingale version:

$$\dot{X}_t = \frac{1}{2}\Delta X_t + \dot{M}_t, \qquad (6.16)$$

where M_t is a martingale measure with $M_0 = 0$ and with quadratic variation

$$\langle M(\phi)\rangle_t = \int_0^t \langle \phi^2, X_s\rangle \, ds.$$

The Eq. (6.16) has two expressions on the right-hand side. The first expression ΔX_t is the deterministic part of the process and accounts for the movement of the cells. If we ignore the birth and death process, we obtain a deterministic process whose values are the densities of the spatial distribution of cells. These densities are solutions of the heat equation $\partial_t f(x, t) = \frac{1}{2}\Delta f(x, t)$. The second expression $\dot{M}_t$ is stochastic and describes fluctuations of the number of cells as a result of cell division and death.

Trajectories of a process (X_t) are measures whose topological supports have interesting geometric properties. Let us denote by $\dim A$ the Hausdorff dimension of the set A. Then

$$\dim \operatorname{supp} X_t = \min(2, d)$$

with probability one. In particular, if $d = 3$, then phytoplankton cells form sets with a fractal structure, although Hausdorff dimension of the support is an integer. Since Brownian motion is spatially symmetric, and the branching process does not depend on space, so the superprocess (X_t) is also spatially symmetric. Supports of measure X_t should have a similar property, but in the three-dimensional space spatially symmetric sets have topological dimension three or zero. This means that typical supports of measures X_t have different topological and Hausdorff dimensions, and therefore they are fractal sets. Since phytoplankton cells form cluster structures similar to fractal sets, the question arises whether the mechanisms proposed in the model can lead in reality to such structures? It is worth noting that the model under consideration was created by a limit passage with the intensity of division and death going to infinity. The paper [94] presents the results concerning the displacement of cells by diffusion as a function of their size. The spatial distribution of cells has a clustered structure if the displacement of cells between divisions is not too large in comparison with the cell size, and this behaviour is observed in the case of relatively large phytoplankton cells.

6.5 Model with Interactions

Some species of phytoplankton secrete substances that attract other individuals of the species. These chemical signals help the phytoplankton cells to move towards the highest concentration of food and avoid polluted areas. Models describing the movement of cells under the influence of chemotaxis have been studied in many publications. We will now present a model based on [1, 39, 94]. The starting point is the system of second order stochastic differential equations which expresses the forces acting on particles in fluid: friction, interaction forces and random collisions (diffusion):

$$\frac{d^2\mathbf{X}_t^i}{dt^2} = -\beta \frac{d\mathbf{X}_t^i}{dt} + \sum_{j=1, j\neq i}^{N} m^{-1}\mathbf{F}(\mathbf{X}_t^i - \mathbf{X}_t^j) + \beta\sqrt{2D}\frac{d\mathbf{W}_t^i}{dt} \tag{6.17}$$

for $i = 1, \ldots, N$, where $\mathbf{X}_t^i$ is the position of the i-th cell at time t, β is the friction constant, m is the mass of a single cell, $\mathbf{F}$ is a pairwise attraction force, D is the diffusion constant and $(\mathbf{W}_t^i)$ is a sequence of N independent three-dimensional Wiener processes. If we consider the movement of particles in fluid we can neglect acceleration because the friction constant β is large in comparison with acceleration (Smoluchowski-Kramers approximation [100]). It means that we can replace system (6.17) by a simpler one:

$$d\mathbf{X}_t^i = \sum_{j=1, j\neq i}^{N} \mathbf{K}(\mathbf{X}_t^i - \mathbf{X}_t^j)\,dt + \sqrt{2D}\,d\mathbf{W}_t^i, \tag{6.18}$$

where $\mathbf{K} = \beta^{-1}m^{-1}\mathbf{F}$. We assume that the attraction forces act similarly to gravitational forces, so they are directed along a straight line connecting the cells, and their magnitude depends only on the distance between the cells $\|\mathbf{x} - \mathbf{y}\|$, hence $\mathbf{K}(\mathbf{z}) = -\frac{\mathbf{z}}{\|\mathbf{z}\|}\varphi(\|\mathbf{z}\|)$, where φ is the length of the vector $\mathbf{K}$. Since cells have a certain size, so the attractive forces cannot act at small distances. In [39] it was assumed that $\varphi(x) = a(x - x_1)(x_2 - x)$ for $x \in [x_1, x_2]$, where $a > 0$ and $x_2 > x_1 > 0$, and $\varphi(x) = 0$ outside of this interval. Despite the advantage associated with the simplicity of this formula, this choice of φ seems unrealistic. In [94], it was proposed that φ depends on the concentration or gradient change of some substance emitted by cells. A simple calculation leads to the formula $\varphi(x) = Cx^{-2}$, when we assume that φ depends on the concentration and $\varphi(x) = Cx^{-3}$ if it depends on gradient change. Of course, these formulas are only satisfied outside the cell.

We now assume that the mass of a single cell is proportional to $1/N$. Then the measure $\xi_t^N = \frac{1}{N}\sum_{i=1}^{N}\delta_{\mathbf{X}_t^i}$ describes the spatial distribution of phytoplankton at

6.5 Model with Interactions

time t. A sequence of processes (ξ_t^N) is weakly convergent to a deterministic process (ξ_t) with values in a space of measures on $\mathbb{R}^3$. This process satisfies the equation

$$\dot{\xi}_t = D\Delta\xi_t - \text{div}((\mathbf{K} * \xi_t)\xi_t), \tag{6.19}$$

where $\mathbf{K} * \xi_t$ is the convolution of the function $\mathbf{K}$ and the measure ξ_t. Equation (6.19) was written in a weak form, i.e. it is satisfied if we have

$$\frac{d}{dt}\langle \psi, \xi_t \rangle = D\langle \Delta\psi, \xi_t \rangle + \langle (\mathbf{K} * \xi_t) \cdot \text{grad } \psi, \xi_t \rangle \tag{6.20}$$

for each test function $\psi \in C_c^\infty$. The symbol C_c^∞ denotes the space of functions of class C^∞ with compact supports. If the measure ξ_0 has the density $u_0(\mathbf{x})$, then also the measure ξ_t has the density $u(t, \mathbf{x})$ satisfying the nonlinear parabolic equation:

$$\frac{\partial}{\partial t}u(t, \mathbf{x}) = D\Delta u(t, \mathbf{x}) - \text{div}((\mathbf{K} * u)(t, \mathbf{x})u(t, \mathbf{x})) \tag{6.21}$$

with the initial condition $u(0, \mathbf{x}) = u_0(\mathbf{x})$, and the convolution $\mathbf{K} * u$ is with respect to the variable $\mathbf{x}$. We obtain the Fokker-Planck equation with the drift coefficient $(\mathbf{K} * u)(t, \mathbf{x})$. The term $(\mathbf{K} * u)(t, \mathbf{x})$ describes the local velocity of phytoplankton which results from the interaction among cells.

In the model considered in the work [39], in addition to cell interactions, birth and death processes are added. A sequence (ξ_t^N) of processes modified in this way converges to a process (ξ_t) satisfying the stochastic partial equation

$$\dot{\xi}_t = D\Delta\xi_t - \text{div}((\mathbf{K} * \xi_t)\xi_t) + \dot{M}_t. \tag{6.22}$$

The authors also present results of computer simulations in the case of $d = 1$ and $d = 2$ concerning spatial structure based on the Clark–Evans index. These simulations show that the introduction of attraction between phytoplankton cells leads to cluster structures.

Chapter 7
Chemotaxis Models

In some models, individuals interact via some chemical factors. Such models are described by IBMs coupled to partial differential equations. We present two models of this kind: the first—motivating, strictly biological—a model of retinal angiogenesis. And the second, conceptually simpler but preserving most mathematical features, is a model of proliferating cells that undergo chemotaxis.

7.1 Model of Retinal Angiogenesis

The model was presented in [26] and it is based on previous models of retinal angiogenesis by Capasso et al. [25], Aubert et al. [6] and McDougall et al. [76], with some ideas taken from tumor angiogenesis models such as in Capasso and Morale [24], Rejniak and Anderson [89], Capasso and Flandoli [23].

Understanding the biological principles that govern blood vessel growth in the retina has important clinical implications, for the prevention of possible retinopathies, which may eventually lead to blindness. Unfortunately, a realistic mathematical model of the process of retinal vasculature, based on current biological knowledge, would lead to a complexity that makes it hard for a sound mathematical analysis. A typical real experimental system is the mouse retina. During embryonic periods, the mouse retina has no proper vascular system, and all gas and material exchange is carried out by diffusion from the hyaloid and choroidal vascular system. Immediately after birth, the retinal vascular system starts to develop, and during the first postnatal week, retinal vessels extend radially over the superficial layer of the retina to form a two-dimensional vascular network. Afterwards, vessels start to make a more complex three-layered system. The model describes only the evolution during the first week, when a two dimensional vascular structure is formed.

The movement and differentiation of cells depend on the concentration of oxygen and some biochemical factors produced by cells in tissue; the most important one is a vascular endothelial growth factor (VEGF) expressed by astrocytes. VEGF stimulates sprouting of new vessels from existing ones; the new vessels are led by the special endothelial tip cells, which are capable of responding to attractant and repellent chemical factors. While moving ahead, tip cells produce mural (stalk) cells, which are eventually responsible for the building up of blood vessels. Blood vessels carry oxygen to the astrocytes, thus inducing negative feedback on the production of further VEGF. Such negative feedback is responsible for establishing a gradient in the concentration of VEGF; tip cells then move forward because of their attractive response to gradients of VEGF.

We consider two types of cells: type 1—*mural cells*—the mature endothelial cells that form vessels and supply nutrient, in particular oxygen, and type 2 cells—endothelial tip cells—that are specialized cells that lead the vascular sprouts. We assume that only tip cells move, and their movement is guided by attracting and repelling chemical factors. Movement of cells is usually modelled by means of Brownian motion with some drift, but we consider here a Newtonian movement with big friction and random changes of velocity, which results in a Langevin-like equation for cells. We include two chemical factors, namely, g—VEGF (*vascular endothelial growth factor*) and u—the relevant nutrient, i.e. *oxygen*, where $g(t, x)$ and $u(t, x)$ denote their concentration at the time t and the point x.

Let us denote by $N_1(t)$ the random number of mural cells that are alive at time t (we assume mural cells do not die) and by $N_2(t)$ the random number of tip cells however born up to time t. As before, from the beginning we construct a sequence of models indexed by the initial number of cells $N = N_1(0) + N_2(0)$. Let us then denote by $(\mathbf{X}_i^{2,N}(t), \mathbf{V}_i^{2,N}(t))$ with $i = 1, \ldots, N_2(t)$ positions and velocities of tip cells (enumerated in an arbitrary order). Given the concentrations u_N and g_N of the VEGF and of oxygen, respectively, the movement of tip cells is then given by

$$\begin{cases} d\mathbf{X}_i^{2,N}(t) = \mathbf{V}_i^{2,N}(t)dt \\ d\mathbf{V}_i^{2,N}(t) = [-k\,\mathbf{V}_i^{2,N}(t) + F(g_N, \nabla g_N, u_N, \nabla u_N)]dt + \sigma dW_i(t), \end{cases} \quad (7.1)$$

where k is a friction coefficient, σ is a strength of Brownian noise, and $W_i(t)$ are independent Brownian motions. The function $F(g_N, \nabla g_N, u_N, \nabla u_N)$ describes the chemotaxic force due to gradients of concentrations of nutrient and VEGF. Mural cells do not move, but each mural cell keeps track of the velocity of the relevant tip cell at the time of production (their number increases because of mutation of tip cells), so we know the 'orientation' of a mural cell, which means the local direction of a vessel. Let us denote by $N_1(t)$ the random number of mural cells that are alive at time t; we will denote by $(\mathbf{X}_i^{1,N}(t), \mathbf{V}_i^{1,N}(t))$ with $i \in 1, \ldots, N_1(t)$, the positions and orientations of murals existing at time t.

We introduce the empirical measures $\xi_N^{[1]}$ and $\xi_N^{[2]}$ of murals and tip cells, respectively. Considering that mural cells, once generated, do not die, and denoting by T_i^2 and Θ_i^2 the random times of birth and death, respectively, of the $i-th$ tip

7.1 Model of Retinal Angiogenesis

cell, we have

$$\xi_N^{[1]}(t)(d\mathbf{x}, d\mathbf{v}) = \frac{1}{N} \sum_{i=1}^{N_1(t)} \delta_{(\mathbf{X}_i^{1,N}(t), \mathbf{V}_i^{1,N}(t))}(d\mathbf{x}, d\mathbf{v}) \in \mathcal{N}_N,$$

$$\xi_N^{[2]}(t)(d\mathbf{x}, d\mathbf{v}) = \frac{1}{N} \sum_{i=1}^{N_2(t)} 1_{t \in [T_i^2, \Theta_i^2)} \delta_{(\mathbf{X}_i^{2,N}(t), \mathbf{V}_i^{2,N}(t))}(d\mathbf{x}, d\mathbf{v}) \in \mathcal{N}_N,$$

where $\mathcal{N}_N \subset \mathcal{M} = \mathcal{M}(\mathbb{R}^2 \times \mathbb{R}^2)$ is the space of measures on the space of positions and velocities.

The oxygen concentration evolves according to the following non-linear PDE:

$$\frac{\partial u_N(t, \mathbf{x})}{\partial t} = -d_u u_N(t, \mathbf{x}) + D_u \Delta u_N(t, \mathbf{x}) + \alpha_u \eta(t, \mathbf{x}, \xi_N^{[1]}(t)), \quad (7.2)$$

and the evolution of VEGF is described by

$$\frac{\partial g_N(t, \mathbf{x})}{\partial t} = -d_g g_N(t, \mathbf{x}) + D_g \Delta g_N(t, \mathbf{x}) + S(u_N(t, \mathbf{x}))$$
$$- \alpha_g g_N(t, \mathbf{x}) \eta(t, \mathbf{x}, \xi_N^{[1]}(t)), \quad (7.3)$$

where D_u and D_g are respective diffusion rates, d_u and d_g are degradation rates, α_u is the supply rate of oxygen, and α_g is the binding rate of VEGF. The function $S(u) = \lambda \max\{\bar{u} - u, 0\}$ describes the production of VEGF by cells; it decreases with respect to the oxygen concentration. λ is the maximal rate in absence of oxygen; for oxygen concentration above the border value $\bar{u}$, no VEGF is produced. Function η denotes the convolution of $\xi_N^{[1]}$ with some mollifying kernel $\kappa : \mathbb{R}^2 \to \mathbb{R}_+$:

$$\eta(t, \mathbf{x}, \xi_N^{[1]}(t)) := \int_{\mathbb{R}^2} \int_{\mathbb{R}^2} \kappa(\mathbf{z} - \mathbf{x}) \xi_N^{[1]}(t)(d\mathbf{z}, d\mathbf{v}). \quad (7.4)$$

The population dynamics of cells consists of four phenomena:

(i) Vessel branching. Vessel tips may emerge from any mural cell (any point of the existing vessels). The probability rate of branching of a single mural cell at position $\mathbf{x}$ at time t depends on the VEGF concentration and is given by $\beta(g_N(t, \mathbf{x}))$; we assume that β is a non-negative smooth function that is bounded with bounded derivatives.
(ii) Death of tip cells. A tip cell dies when the vessel growth stops, and this event is connected with the sufficient oxygenation of the tissue when new vessels are no longer necessary. So the death rate of tip cells $\mu = \mu(u)$ depends on the local concentration of oxygen at the position of the cell.
(iii) Production of mural cells. The vessels are built on the trajectories of tip cells. In real angiogenesis, the main production of mural cells takes place behind

the leading tip cell. The model simplifies this phenomenon by assuming that mural cells are produced and left behind moving mural cells. Since tip cells move with random velocity, to assure that the linear density of mural cells in vessels is statistically constant, we assume that the production rate depends on the value of tip velocity and is given by $m_{21}|\mathbf{v}|$. The term $m_{21}|\mathbf{v}|$ expresses the fact that the birth of a mural cell is induced by the elongation $|\mathbf{v}|\,ds$ of the trajectory of a tip cell located at $\mathbf{x}$, and having velocity $\mathbf{v}$, at time s, during the time interval $(s, s+ds]$.

(iv) Proliferation of mural cells is an additional mechanism of production of mural cells, and the individual rate of proliferation of a mural cell is given by a constant β_1. After a division of a mural cell, a new one is produced at a random position given by a probabilistic kernel κ_2 around the position of a mother cell. We assume that the new cell keeps memory of the velocity of the mother cell. This mechanism describes thickening of existing vessels.

In [26] the numerical simulations of the described model were conducted. The results are presented in Fig. 7.1. They show that the model can somehow reproduce the complicated branching structure of the vascular plexus.

The main difference between the model of retinal angiogenesis and the models presented in the previous chapters is that the movement and proliferation of cells depend on the fields governed by PDEs. It means that the IBM is strongly coupled to PDEs in the sense that the behaviour of the IBM depends on the solutions of the PDE and the right-hand side of the PDE depends on the position of particles of the IBM. The number of endothelial cells is large, especially after some time, and similarly as before, the question arises: what is the large number of particles limit. In the paper [26] the limit PDEs were heuristically derived to be of the form

$$\partial_t \rho_1(t,\mathbf{x},\mathbf{v}) = m_{21}|\mathbf{v}|\,\rho_2(t,\mathbf{x},\mathbf{v}) + \beta_1 \int_{\mathbb{R}^2} \kappa_2(\mathbf{z}-\mathbf{x})\rho_1(t,\mathbf{z},\mathbf{v})\,d\mathbf{z},$$

$$\partial_t \rho_2(t,\mathbf{x},\mathbf{v}) = \frac{\sigma^2}{2}\Delta_\mathbf{v}\rho_2(t,\mathbf{x},\mathbf{v}) - \mathbf{v}\cdot\nabla_\mathbf{x}\rho_2(t,\mathbf{x},\mathbf{v})$$
$$-\operatorname{div}_\mathbf{v}\left(\left[-k\mathbf{v} + F(g(t,\mathbf{x}), \nabla g(t,\mathbf{x}), u(t,\mathbf{x}), \nabla u(t,\mathbf{x}))\right]\rho_2(t,\mathbf{x},\mathbf{v})\right)$$
$$+\beta(g(s,\mathbf{x}))\int_{\mathbb{R}^2} G_\mathbf{z}(\mathbf{v})\,\rho_1(t,\mathbf{x},\mathbf{z})d\mathbf{z} - \mu(u(s,\mathbf{x}))\rho_2(t,\mathbf{x},\mathbf{v}),$$

coupled to

$$\frac{\partial u(t,\mathbf{x})}{\partial t} = -d_u u(t,\mathbf{x}) + D_u \Delta u(t,\mathbf{x}) + \alpha_u \tilde{\eta}(t,\mathbf{x}),$$

$$\frac{\partial g(t,\mathbf{x})}{\partial t} = -d_g g(t,\mathbf{x}) + D_g \Delta g(t,\mathbf{x}) + S(u(t,\mathbf{x})) - \alpha_g g(t,\mathbf{x})\tilde{\eta}(t,\mathbf{x}),$$

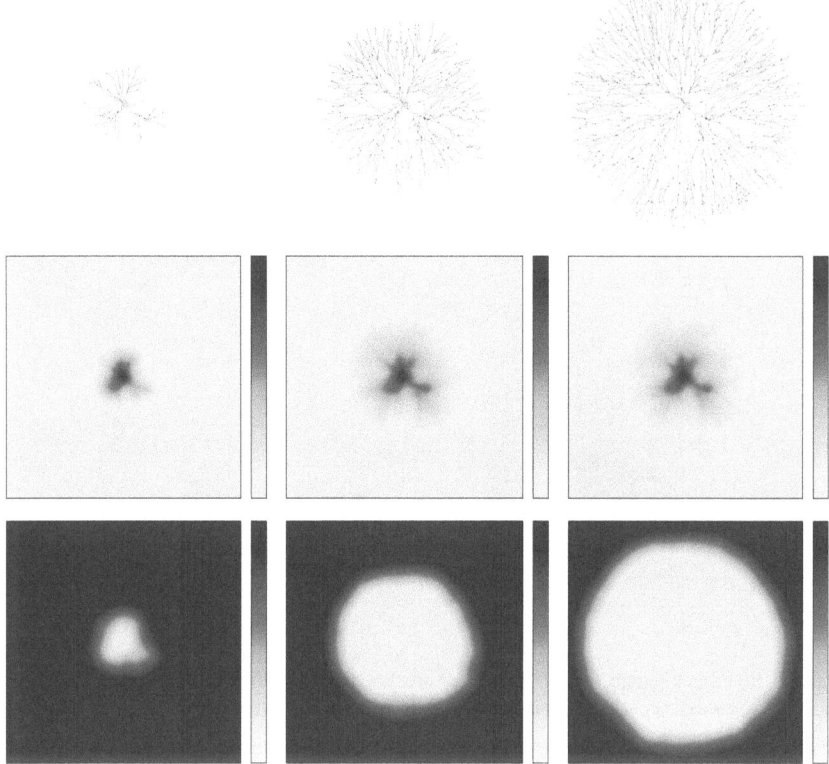

Fig. 7.1 Sequence of snapshots showing the simulated development of the capillary structure. the upper row: snapshots of the vessel network at three different times; the middle row: density plots of the oxygen concentration; the lower row: density plots of the growth factor concentration. The figure from: Capasso, Wieczorek [26]

with

$$\tilde{\eta}(t, \mathbf{x}) := \int_{\mathbb{R}^2} \int_{\mathbb{R}^2} \kappa(\mathbf{z} - \mathbf{x}) \rho_1(t, \mathbf{z}, \mathbf{v}) \, d\mathbf{z} \, d\mathbf{v}.$$

7.2 Branching Diffusion with Location Dependent Branching Rates

For better understanding of the nature of the convergence in the previous section, and in fact, in previous chapters, we present here a sequence of simple processes living on $\mathbb{R}^2$ with some pictures.

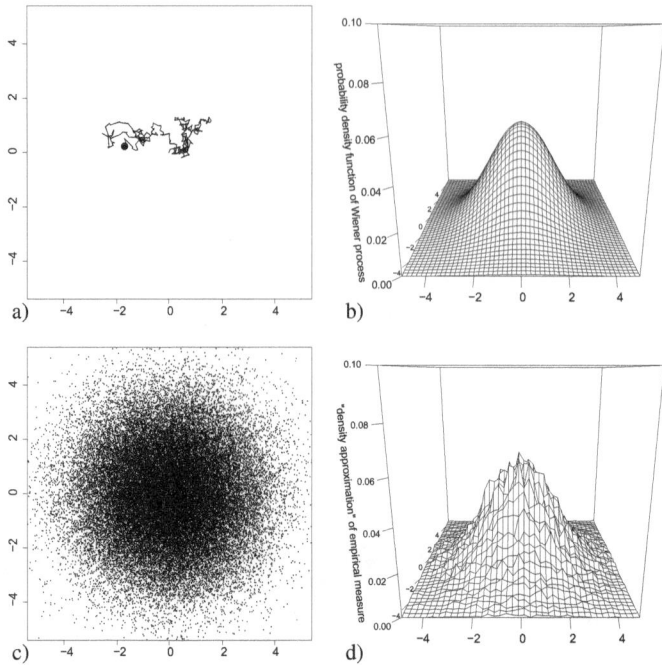

Fig. 7.2 (**a**) a sample trajectory of the Wiener process; (**b**) the probability distribution function of the Wiener process; (**c**) sample positions of 50000 independent Wiener processes; (**d**) a histogram of points in (**c**)

Firstly, let us look at the sample trajectory up to time 3 in Fig. 7.2a and a probability distribution function at time 3 of the standard Wiener process on $\mathbb{R}^2$ in Fig. 7.2b. Next, take a large number N of independent Wiener processes W_t^i, and consider their empirical distribution $\bar{\xi}_t^N = \frac{1}{N}\sum_{i=1} \delta_{W_t^i}$. Figure 7.2c presents the sample empirical distribution—each dot represents a position of one Wiener process, while in Fig. 7.2d we see the "approximated density" of this empirical measure: the histogram of points from Fig. 7.2c. Surely, Fig. 7.2d is very similar to Fig. 7.2b and the convergence follows from the simple classical law of large numbers. Namely, for a fixed $t > 0$ and a Borel set $B \subset \mathbb{R}^2$

$$\bar{\xi}_t^N(B) = \frac{1}{N}\#\{i \in \{1,\ldots,N\} : W_t^i \in B\} = \frac{1}{N}\sum_{i=1} \delta_{W_t^i}(B)$$

by the law of large numbers, converges to $\mathrm{E}\,\delta_{W_t^1}(B) = \mathrm{E}\,1_B(W_t^1) = \mathrm{P}(W_t^1 \in B)$.

Now, let us add birth and death to the model. Take a single Brownian particle that after exponentially distributed time either dies or duplicates, creating two independent Brownian particles. Each of these particles moves independently and has its own exponential clock. Then it either dies or duplicates again. Such a

7.2 Branching Diffusion with Location Dependent Branching Rates

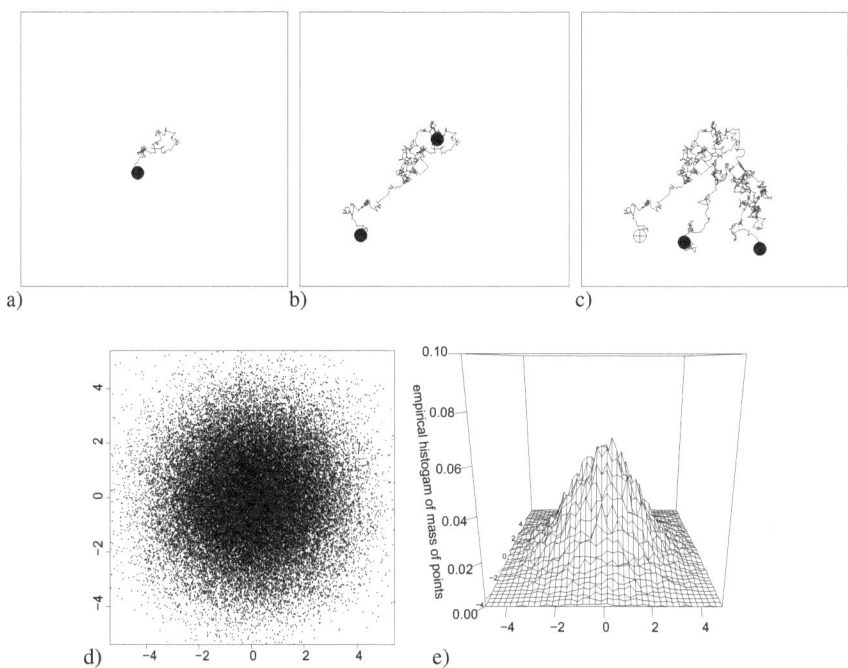

Fig. 7.3 In the first row, three snapshots of a single trajectory of a branching diffusion starting from a single particle: (**a**) diffusion of the first particle (• marks positions of living particles); (**b**) after first branching (⋄ is the point of branching); (**c**) after second division (⋄) and first death (⊕). Second row: (**d**) positions of all living particles of 50,000 independent branching diffusions after time 3; (**e**) a histogram of points from (**d**)

process is called *branching Brownian motion* or more general *branching diffusion* (cf. Chap. 4). The first row of Fig. 7.3 shows three snapshots of a single trajectory of a branching Brownian motion on the plane. Note that this process is no more $\mathbb{R}^2$-valued, since the number of particles changes in time. There are a few possible spaces to describe it, but the most common one is the space of integer-valued measures $\mathcal{N} = \left\{ \sum_{i=1}^{k} \delta_{x_i} : k \in \mathbb{N}, \ x_i \in \mathbb{R}^d \right\}$ (cf. Sect. 1.2).

Let us now consider a sequence (ξ_t^i) of N independent branching Brownian motions staring from one particle in the origin and take the average $\bar{\xi}_t^N = \frac{1}{N} \sum_{i=1}^{N} \xi_t^i$. Note that we divide by N but the number of particles in each (ξ_t^i) varies, so $\bar{\xi}_t^N$ is not a probabilistic measure. Nevertheless, if N is large, then $\bar{\xi}_t^N$ looks like the empirical distribution of N Brownian motions, and if we compare Fig. 7.3d, where we have positions of all particles of $\bar{\xi}_t^N$ at time 3, with Fig. 7.2c they are indistinguishable. Similarly, the histogram of all points of $\bar{\xi}_t^N$ in Fig. 7.3e looks exactly like Fig. 7.2d. Obviously, it is not a coincidence: using the law of large numbers for any $B \in \mathcal{B}(\mathbb{R}^2)$, we get that $\bar{\xi}_t^N(B)$ converges to $\mathrm{E}\,\xi_t^1(B)$ and this is equal to $\mathrm{P}(W_t^1 \in B)$, which will be clear from Lemma 7.1.

Consider now a branching Brownian motion with position-dependent birth and death rates, i.e. a system of particles that move according to independent Wiener processes and may die with the rate $d(x)$ or duplicate with the rate $b(x)$, where $b : \mathbb{R}^d \to [0, \infty)$ and $d : \mathbb{R}^d \to [0, \infty)$ are bounded and continuous. Again, there are a couple of ways to describe such a process; one is an $\mathcal{N}$-valued process with the generator

$$LF_h(\nu) = \exp\langle h, \nu\rangle \left\langle \frac{\frac{1}{2}\Delta e^h}{e^h} + \left(b e^h + d e^{-h} - (b+d)\right), \nu \right\rangle, \qquad (7.5)$$

where $F_h(\nu) = \exp\langle h, \nu\rangle$. We will show here the second possible description, which is convenient in calculations and will be useful in the next section. Similarly to the proof of Lemma 3.3, we want to assign to each particle its own Poisson random measure and Wiener process, so we need to distinguish the particles. One can do it as in Lemma 3.3 by introducing some order on $\mathscr{X}$ (cf. [49, 90, 113]), or by explicitly numbering particles by some subspace of the Ulam-Harris tree [43, 60, 110] and remembering the whole family history of the particles. Because we will need the latter approach later, we use it here. We assume that each particle has at most two descendants, so we number all (possible) particles by elements of the set

$$\mathbb{J} = \bigcup_{n=0}^{\infty} \{0, 1\}^n = \{\emptyset\} \cup \{j_1 \ldots j_n : n \in \mathbb{N}_+, \ j_k \in \{0, 1\} \text{ for } k = 1, \ldots, n\},$$
(7.6)

with the convention that $\{0, 1\}^0 = \emptyset$ marks the root (initial) particle (see Fig. 7.4). Now, we assign to each node of the tree $\mathbb{J}$ either a position at $\mathbb{R}^d$ or an external 'cemetery' state $*$ which means that this particle does not exist. So, a state of our particle system is given by an element of the space

$$\mathbb{X} = \left\{(x_\mathbf{j})_{\mathbf{j} \in \mathbb{J}} : x_\mathbf{j} \in \mathbb{R}^d \cup \{*\} \text{ and there is at most a finite number of } \mathbf{j} \text{ s.t. } x_\mathbf{j} \neq *\right\}.$$

Let now $X_\mathbf{j}(t) \in \mathbb{R}^d$ denote the position of $\mathbf{j}$-th particle, $\mathbf{j} \in \mathbb{J}$, if the particle exists at time t, or $X_\mathbf{j}(t) = *$ if the particle has not been born yet or has already

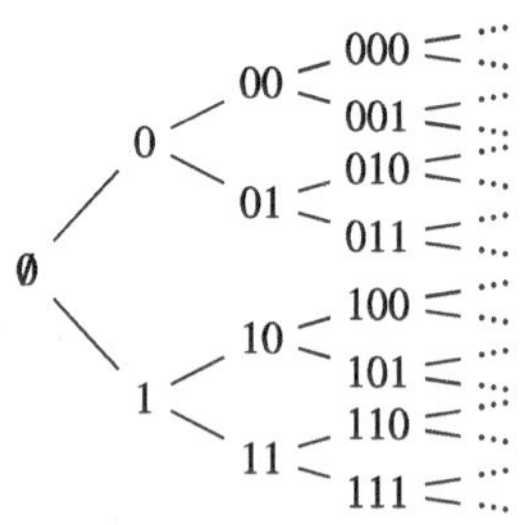

Fig. 7.4 The Ulam-Harris tree $\mathbb{J}$

7.2 Branching Diffusion with Location Dependent Branching Rates

died. We assume that in the moment of branching the old particle, say j-th, dies and either leaves nothing behind or leaves two offspring numbers as j0 and j1. Take an array $(W_j(t))_{j \in \mathbb{J}}$ of independent Brownian motions and an array $(\mathfrak{N}_j(dt,dz))_{j \in \mathbb{J}}$ of independent space-time Poisson random measures with intensity measure $dt\,dz$ on $[0,\infty) \times [0,\infty)$. Then we can describe the branching diffusion process as the solution to the infinite system of Itô-Lévy stochastic differential equations indexed by $j \in \mathbb{J}$

$$dX_j(t) = 1_{\mathbb{R}^d}(X_j(t))\,dW_j(t), \tag{7.7}$$

$$+ \int_0^\infty \chi_b(X_{j^-}^N(t^-),z)\mathfrak{N}_{j^-}(dt,dz) + \int_0^\infty \chi_d(X_j^N(t^-),z)\mathfrak{N}_j(dt,dz),$$

where

$$\chi_b(x,z) = \begin{cases} -*+x, & \text{if } x \neq *,\ z \leq b(x), \\ 0, & \text{otherwise}, \end{cases}$$

$$\chi_d(x,z) = \begin{cases} *, & \text{if } x \neq *,\ z \leq b(x)+d(x), \\ 0, & \text{otherwise}, \end{cases}$$

with the convention that $x + * = *$ and $* - * + x = x$ for any $x \in \mathbb{R}^d$. We have denoted by j^- the index of a parent particle of the j-th particle, i.e., if e.g. $j = 10011$ then $j^- = 1001$. To understand how this equation works, note that if $X_j(t) = *$ then $1_{\mathbb{R}^d}(X_j(t)) = 0$ and nothing changes up to the time t when the parent's clock $\mathfrak{N}_{j^-}$ clicks, i.e. $\mathfrak{N}_{j^-}(\{t,z\}) = 1$ with a z such that $z \leq b(X_{j^-}(t))$ and then $\chi_b(X_{j^-}^N(t^-),z) = -*+X_{j^-}^N(t^-)$, so $X_j(t)$ jumps from $*$ to the parent's position $X_{j^-}^N(t^-)$. At the same time in the equation for the parent $i = j^-$ the term $\chi_d(X_i^N(t^-),z)\mathfrak{N}_i(dt,dz)$ works and the parent dies, jumping from $X_{j^-}^N(t^-)$ to $*$. On the other side, when $X_j(t) \neq *$ the term $1_{\mathbb{R}^d}(X_j(t))\,dW_j(t)$ results in movement according to $W_j(t)$, function $\chi_b(X_{j^-}^N(t^-),z)$ is always 0, since the parent j^- is already dead. Moreover, when there appears a point (t,z) such that $\mathfrak{N}_j(\{t,z\}) = 1$ and $z \leq b(X_j(t)) + d(X_j(t))$ then the last term of (7.7) works with $\chi_d(X_j^N(t^-),z) = *$, so by the convention the process jumps from $X_j^N(t^-)$ to $*$, which means death. At the same time, if $z \leq b(X_j(t))$, then in two equations describing the descendants of j, namely $j0$ and $j1$, the birth term clicks (note that $j = j0^- = j1^-$) and the $j0$-th and $j1$-th processes jump form $*$ to $X_j^N(t^-)$, while if $b(X_j(t)) < z \leq b(X_j(t)) + d(X_j(t))$ nothing happens because $\chi_b(X_{j0^-}^N(t^-),z) = \chi_b(X_{j1^-}^N(t^-),z) = 0$.

Note that, since the total number of particles is less (in distribution) than a pure birth process with a birth rate of $\sup_{x \in \mathbb{R}^d} b(x)$ (cf. Sect. 3.2 and Lemma 3.1), the total number of alive particles is finite with probability one, so there is actually always a finite number of equations in (7.7), and they are well defined.

We can define the measure-valued process by

$$\xi_t = \sum_{\substack{j \in \mathbb{J} \\ X_{\mathbf{j}}(t) \neq *}} \delta_{X_{\mathbf{j}}(t)} \tag{7.8}$$

and check that its generator is (7.5). Now we may use the properties of Itô-Lévy integral to check that

Lemma 7.1 *Let ξ_t be a measure-valued process defined by (7.8), where $X_{\mathbf{j}}(t)$ are solutions of (7.7) and let $E\langle 1, \xi_0\rangle < \infty$. Assume that $b, d \in C_b^2(\mathbb{R}^d)$. The measure $\bar{\mu}_t(B) = E \xi_t(B) = E\#\{\mathbf{j} : X_{\mathbf{j}} \neq *\}$, i.e. the expected number of particles in a set $B \in \mathscr{B}(\mathbb{R}^d)$, is the unique solution starting from $\bar{\mu}_0 = E\xi_0$ to the following problem*

$$\frac{d\langle h, \bar{\mu}_t\rangle}{dt} = \left\langle \tfrac{1}{2}\Delta h(\cdot) + \big(b(\cdot) - d(\cdot)\big) h(\cdot), \bar{\mu}_t \right\rangle \qquad \text{for all } h \in C_b^2(\mathbb{R}^d). \tag{7.9}$$

Proof Note that $\langle h, \bar{\mu}_t\rangle = \langle h, E\xi_t\rangle = E\langle h, \xi_t\rangle = E\left[\sum_{\mathbf{j}} h(X_{\mathbf{j}}(t))\right]$. By the Itô-Lévy formula let us calculate

$$\begin{aligned} d\left[h(X_{\mathbf{j}}(t))\right] =& \nabla h(X_{\mathbf{j}}(t))\, dW_{\mathbf{j}}(t) + \tfrac{1}{2}\Delta h(X_{\mathbf{j}}(t))\, dt \\ &+ h(X_{\mathbf{j}^-}(t^-)) 1_{\left[0, b\left(X_{\mathbf{j}^-}(t^-)\right)\right]}(z)\, \mathfrak{N}_{\mathbf{j}^-}(dt, dz) \\ &- h(X_{\mathbf{j}}(t^-)) 1_{[0, b(X_{\mathbf{j}}(t^-)) + d(X_{\mathbf{j}}(t^-))]}(z)\, \mathfrak{N}_{\mathbf{j}}(dt, dz). \end{aligned}$$

Taking expectation, by the compensator formula we get

$$\begin{aligned} d\, E\left[h(X_{\mathbf{j}}(t))\right] =& E\left[\tfrac{1}{2}\Delta h(X_{\mathbf{j}}(t))\right] dt + E\, h(X_{\mathbf{j}^-}(t)) b(X_{\mathbf{j}^-}(t))\, dt \\ &- E\, h(X_{\mathbf{j}}(t)) \big(b(X_{\mathbf{j}}(t)) + d(X_{\mathbf{j}}(t))\big)\, dt. \end{aligned}$$

Summing over $\mathbf{j} \in \mathbb{J}$ and renumerating to get rid of $\mathbf{j}^-$ we have

$$\begin{aligned} d\, E \sum_{\mathbf{j} \in \mathbb{J}} \left[h(X_{\mathbf{j}}(t))\right] =& E \sum_{\mathbf{j} \in \mathbb{J}} \left[\tfrac{1}{2}\Delta h(X_{\mathbf{j}}(t))\right] dt \\ &+ E \sum_{\mathbf{j} \in \mathbb{J}} \big(b(X_{\mathbf{j}}(t)) - d(X_{\mathbf{j}}(t))\big) h(X_{\mathbf{j}}(t))\, dt \\ =& \left\langle \tfrac{1}{2}\Delta h(\cdot) + \big(b(\cdot) - d(\cdot)\big) h(\cdot), \bar{\mu}_t \right\rangle dt, \end{aligned}$$

7.2 Branching Diffusion with Location Dependent Branching Rates

which proves that $\bar{\mu}_t$ satisfies (7.9). To prove the uniqueness, the duality argument is used. Namely, let $\bar{\mu}_t$ satisfy (7.9), take $h \in C_c^2(\mathbb{R}^d)$ and let $u(t,x)$ be a solution to the equation

$$\frac{\partial u(t,x)}{\partial t} = \tfrac{1}{2}\Delta u(t,x) + (b(x) - d(x))u(t,x) \qquad (7.10)$$

with $u(0,x) = h(x)$. Using (7.10) and (7.9) for $s \in [0,t]$ we get that

$$\frac{d}{ds}\langle u(t-s,\cdot), \bar{\mu}_s \rangle = 0,$$

so integrating on $[0,t]$ we have

$$\langle h, \bar{\mu}_t \rangle = \langle u(t,\cdot), \bar{\mu}_0 \rangle.$$

Therefore, $\langle h, \bar{\mu}_t \rangle$ is uniquely defined for any $h \in C_b^2(\mathbb{R}^d)$ which proves that $\bar{\mu}_t$ is unique. □

Note that (7.9) is a weak version of the diffusion equation with growth

$$\frac{\partial u(t,x)}{\partial t} = \tfrac{1}{2}\Delta u(t,x) + (b(x) - d(x))u(t,x). \qquad (7.11)$$

Although it looks exactly like (7.10), they are actually adjoint and they would be different if we considered a more general diffusion and an eliptic operator instead of the Laplacian. Additionally, we have proven uniqueness of measure-valued solutions to (7.9), and (7.11) has a smooth fundamental solution $\Phi(t,x)$, so $\bar{\mu}_t = \int_{\mathbb{R}^d} \Phi(t,x)\bar{\mu}_0(dx)$ is also smooth. Moreover, note that if $b = d$ as in Fig. 7.3, then $\bar{\mu}_t$ is just a solution of the heat equation and that justifies why Fig. 7.3e is so similar to Fig. 7.2d and b. In Fig. 7.5 we have illustrated the law of large numbers

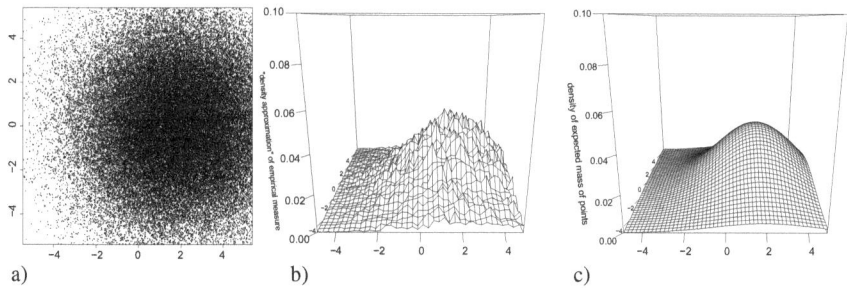

Fig. 7.5 Position-dependent branching diffusion. The birth rate b is greater on the right-hand side while the death rate d is greater on the left-hand side. (**a**) positions of all living particles of 50,000 independent branching diffusions; (**b**) a histogram of points from (**a**); (**c**) a solution to the diffusion equation with growth (7.11)

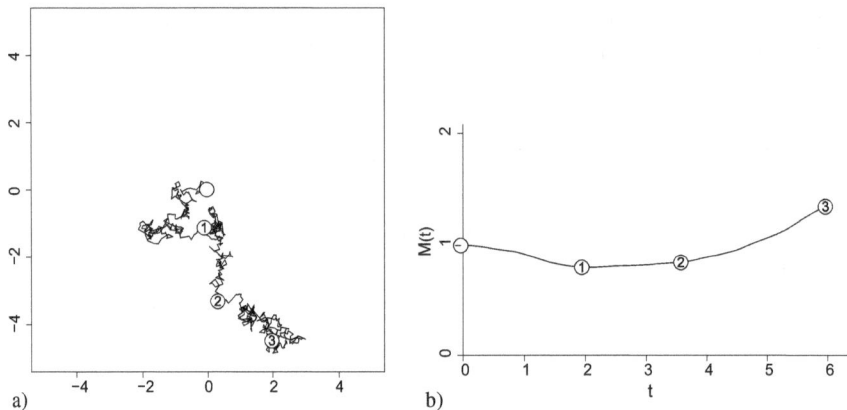

Fig. 7.6 A Brownian particle with mass: (**a**) sample trajectory of Brownian particle X_t on the plane; (**b**) sample trajectory of mass M_t of the particle vs. time. As before, $b - d$ is negative at the left-hand side and positive at the right-hand side. While the particle stays on the left (from ○ to ①), the mass $M(t)$ decreases, between ① and ② the particle stays near the middle and the mass stays approximately constant, between ② and ③ the particle is on the right-hand side and its mass grows

for position-dependent branching diffusions, by comparing the histogram of 50,000 independent branching diffusions with the solution of (7.11).

We give here one more interpretation of (7.9) or (7.11). Consider a single Brownian particle (X_t) and let us assign to it a mass (M_t), that changes according to the following equation:

$$dM_t = (b(X_t) - d(X_t))\,dt,$$

see Fig. 7.6. Then the measure describing the expected mass of this particle in given set, i.e. $\tilde{\mu}(B) = \mathrm{E}\,M_t \delta_B(X_t)$ satisfies (7.9). Indeed, $\langle h, \tilde{\mu}(B)\rangle = \mathrm{E}\,M_t h(X_t)$ and by the Itô formula we have

$$d\,[M_t h(X_t)] = M_t\left(\nabla h(X_t) \cdot dW_t + \tfrac{1}{2}\Delta h(X_t)\,dt\right) + [b(X_t) - d(X_t)]M_t h(X_t)\,dt$$

and taking expectation we get

$$d\,[\mathrm{E}\,M_t h(X_t)] = \left[\tfrac{1}{2}\Delta h(X_t) + (b(X_t) - d(X_t))h(X_t)\right] M_t\,dt$$
$$= \left\langle \tfrac{1}{2}\Delta h(\cdot) + \big(b(\cdot) - d(\cdot)\big)h(\cdot), \tilde{\mu}_t\right\rangle dt.$$

7.3 Model of Proliferating Cells with Chemotaxis

Based on the previous section, we present a model, introduced in [110], that is simpler than the one from Sect. 7.1, but still is multiscale, i.e., consists of IBM coupled to PDEs and includes both chemotaxis and birth and death processes. We will show the law of large number-type theorem, which may be called in this case the mean-field approximation, that will lead to a nonlinear model.

Let us consider a population of cells that proliferate and move, and the movement and proliferation are regulated by a chemoreactant produced by cells themselves. We construct a sequence of processes indexed by the initial number of particles N. The description of our process can be divided into three components:

Movement of Cells We consider N initial cells located at $X_{i,\emptyset}^N(0)$, where $\emptyset$ means root of the Ulam-Harris tree $\mathbb{J}$, see Fig. 7.4. We mark the possible cell by the number $i \in 1, \ldots, N$ which means the number of a 'cell line', i.e. the number of the oldest ancestor—one of the initial cells, and, as in the previous section, by the element of $\mathbb{J}$, that is the history of all ancestors from the $(i, \emptyset)$-th. The cells move according to the following SDE

$$dX_{i,\mathbf{j}}^N(t) = \mathbf{c}(X_{i,\mathbf{j}}^N(t), \nabla \rho_N(t, X_{i,\mathbf{j}}^N(t)))\, dt \\ + \sigma\, dW_{i,\mathbf{j}}(t), \text{ for } i = 1, \ldots, N, \text{ and } \mathbf{j} \in \mathbb{J}, \qquad (7.12)$$

where $W_{i,\mathbf{j}}(t)$ are independent Brownian motions, σ is the diffusion coefficient and $\mathbf{c}$ is the (chemotactic) drift that depends on the position of a cell and the gradient of the concentration ρ_N of some chemical factor. One can also allow for the dependence of $\mathbf{c}$ on the concentration ρ_N itself, not only on its gradient, but for the sake of shortness and simplicity of the notation this dependence is neglected.

Equation for Chemoreactant The concentration ρ_N of chemotactic factor satisfies the following PDE

$$\frac{\partial \rho_N(t, X)}{\partial t} = D\Delta\rho_N(t, X) - r\rho_N(t, X) + \alpha\, \kappa * \xi_t^N(X), \qquad (7.13)$$

where D, r and α are diffusion, degradation and production rates. The measure

$$\xi_t^N = \frac{1}{N} \sum_{\substack{i,\mathbf{j} \\ X_{i,\mathbf{j}} \neq *}} \delta_{X_{i,\mathbf{j}}^N(t)}, \qquad (7.14)$$

is the empirical measure of all cells and the function κ is a mollifying kernel that represents the fact that cells are actually not points, but have spatial size. The spatial convolution

$$\kappa * \xi_t^N(X) = \int_{\mathbb{R}^d} \kappa(x-y)\xi_t^N(dy) = \frac{1}{N} \sum_{\substack{i,j \\ X_{i,j} \neq *}} \int_{\mathbb{R}^d} \kappa(x-y)\delta_{X_{i,j}(t)}(dy)$$

$$= \frac{1}{N} \sum_{\substack{i,j \\ X_{i,j} \neq *}} \kappa(x - X_{i,j}(t))$$

is a mollified version of the empirical measure describing spatial positions of cells, responsible for the production of the chemoreactant. From the mathematical point of view this allows us to consider classical solutions to Eq. (7.13).

Cell Population Dynamics We assume that cells may die or proliferate with rates depending on the chemoreactant. Death means that a cell disappears and proliferation means that a cell dies leaving two new daughter cells at the same place as the mother cell. The birth rate of a cell placed at x at time t depends on the position and on the concentration of chemoreactant and is given by $b(x, \rho_n(t, x))$ and the mortality rate is $d(x, \rho_n(t, x))$.

As in the previous section we can represent the IBM part as the solution to the system of Itô-Lévy stochastic differential equations indexed by $\mathbf{j} \in \mathbb{J}$. Take, as before, an array $\left(W_{i,\mathbf{j}}(t)\right)_{i \in \mathbb{N}, \mathbf{j} \in \mathbb{J}}$ of independent Brownian motions and an array $\left(\mathfrak{N}_{i,\mathbf{j}}(dt, dz)\right)_{i \in \mathbb{N}, \mathbf{j} \in \mathbb{J}}$ of independent space-time Poisson random measures with intensity measure $dt\, dz$ on $[0, \infty) \times [0, \infty)$. Let $\left(X_{i,\mathbf{j}}(t)\right)_{i \in \mathbb{N}, \mathbf{j} \in \mathbb{J}}$ be a solution of

$$X_{i,\mathbf{j}}(t) = \int_0^t \mathbf{c}(X_{i,\mathbf{j}}^N(t), \nabla \rho_N(t, X_{i,\mathbf{j}}^N(t)))\, dt + \int_0^t \mathbf{1}_{\mathbb{R}^d}\left(X_{i,\mathbf{j}}(t)\right) dW_{i,\mathbf{j}}(t),$$
(7.15)

$$+ \int_0^t \int_0^\infty \chi_b^N(X_{i,\mathbf{j}}^N(t^-), z)\mathfrak{N}_{i,\mathbf{j}^-}(dt, dz)$$

$$+ \int_0^t \int_0^\infty \chi_d^N(X_{i,\mathbf{j}}^N(t^-), z)\mathfrak{N}_{i,\mathbf{j}}(dt, dz),$$

where χ_b^N and χ_d^N depend on ρ_N

$$\chi_b^N(x, z) = \begin{cases} -*+x, & \text{if } x \neq *,\ z \leq b(x, \rho_N(t, x)), \\ 0, & \text{otherwise,} \end{cases}$$

$$\chi_d^N(x, z) = \begin{cases} *, & \text{if } x \neq *,\ z \leq b(x, \rho_N(t, x)) + d(x, \rho_N(t, x)), \\ 0, & \text{otherwise,} \end{cases}$$

7.3 Model of Proliferating Cells with Chemotaxis

retaining the convention that $x + * = *$ and $* - * + x = x$ for any $x \in \mathbb{R}^d$ and $\mathtt{j}^-$ is the index of a parent particle of $\mathtt{j}$-th particle. We have written Eq. (7.15) in the integrated form, to stress the fact, that since σ is constant, we can actually solve it along trajectories, knowing that

$$\int_0^t 1_{\mathbb{R}^d}\left(X_{i,\mathtt{j}}(t)\right) dW_{i,\mathtt{j}}(t) = W_{i,\mathtt{j}}(t) - W_{i,\mathtt{j}}(\tau_{i,\mathtt{j}}) \text{ for } t \in [\tau_{i,\mathtt{j}}, \theta_{i,\mathtt{j}}),$$

where $\tau_{i,\mathtt{j}}$ and $\theta_{i,\mathtt{j}}$ denote birth and death time of $(i,\mathtt{j})$-th cell.

Now we present three nonlinear models that are in a sense equivalent, and then we will formulate a convergence theorem.

Macroscopic Model The hydrodynamic (infinite number of particles) limit of the IBM coupled to PDEs may be described as a weak version of the following Patlak-Keller-Segel type system of PDEs with proliferation

$$\begin{aligned}\partial_t u(t,\mathbf{x}) &= \tfrac{1}{2}\Delta u(t,\mathbf{x}) + \nabla\left(u(t,\mathbf{x})\mathbf{c}(\mathbf{x},\nabla\rho)\right) + \lambda(\mathbf{x},\nabla\rho(t,\mathbf{x}))u(t,\mathbf{x}),\\ \partial_t \rho(t,\mathbf{x}) &= D\Delta\rho(t,\mathbf{x}) - r\rho(t,\mathbf{x}) + \alpha[\kappa * u(t,\cdot)](\mathbf{x}),\end{aligned} \qquad (7.16)$$

where $\lambda = b - d$ and $[\kappa * u(t,\cdot)](\mathbf{x}) = \int_{\mathbb{R}^d} \kappa(x-y)u(t,y)\,dy$.

Hybrid Model In the sequence of individual models defined by (7.13) and (7.15) the process starts from N particles, and the PDE on the field (7.13) depends on the empirical measure of the IBM. Now, we define a model in which we still have branching diffusion IBM, but it starts from a single particle and is coupled to the PDE that depends on the expected distribution of mass of the IBM, namely

$$\bar{X}_{\mathtt{j}}(t) = \bar{X}_{\mathtt{j}}(0) + \int_0^t \mathbf{c}(\bar{X}_{\mathtt{j}}(s), \nabla\rho(t,\bar{X}_{\mathtt{j}}(s)))\,ds \qquad (7.17)$$

$$+ \sigma \int_0^t 1_{\mathbb{R}^d}\left(\bar{X}_{\mathtt{j}}(s)\right) dW_{1,\mathtt{j}}(s),$$

$$+ \int_0^t \int_{\mathbb{R}_+} \chi_b(\bar{X}_{\mathtt{j}^-}(s^-), z)\mathfrak{N}_{1,\mathtt{j}^-}(ds,dz)$$

$$+ \int_0^t \int_{\mathbb{R}_+} \chi_d(\bar{X}_{\mathtt{j}}(s^-), z)\mathfrak{N}_{1,\mathtt{j}}(ds,dz),$$

for $\mathtt{j} \in \mathbb{J}$, where χ_b and χ_d are defined like χ_b^N and χ_d^N with ρ_N replaced by ρ, which evolves according to

$$\frac{\partial \rho(t,\mathbf{x})}{\partial t} = D\Delta\rho(t,\mathbf{x}) - r\rho(t,\mathbf{x}) + \alpha[\kappa * \bar{\mu}_t](\mathbf{x}), \qquad (7.18)$$

where $\bar{\mu}_t$ is the expected value of the empirical measure of $(\bar{X}_j(t))_{j \in \mathbb{J}}$

$$\bar{\mu}_t(B) = \mathrm{E}\left[\sum_{j,\, \bar{X}_j \neq *} \delta_{\bar{X}_j(t)}(B)\right], \quad \text{for } B \in \mathscr{B}(\mathbb{R}^d). \tag{7.19}$$

One Particle Model Let us consider a single cell that moves according to the same rule as one of the particles in the previous model, and endow it with a process $M(t)$ denoting its mass:

$$\begin{aligned} dX(t) &= \mathbf{c}(X(t), \nabla \rho(t, X(t)))dt + \sigma\, dW(t), \\ dM(t) &= \lambda\big(X(t), \rho(t, X(t))\big)M(t)\, dt, \end{aligned} \tag{7.20}$$

coupled to a PDE for the field ρ

$$\frac{\partial \rho(t, \mathbf{x})}{\partial t} = D\Delta\rho(t, \mathbf{x}) - r\rho(t, \mathbf{x}) + \alpha[\kappa * \tilde{\mu}_t](\mathbf{x}), \tag{7.21}$$

where

$$\tilde{\mu}_t(B) = \mathrm{E}\left[M(t)1_B(X(t))\right] \text{ for } B \in \mathscr{B}(\mathbb{R}^d)$$

is the *average mass in the area B*. By an argument similar to Lemma 7.1 one can show that the measure $\tilde{\mu}_t$ is equal to $\bar{\mu}_t$. Note that Eqs. (7.18) and (7.21) are then the same.

Now we can state two convergence theorems from [110].

Theorem 7.1 *Assume that*

- *σ, D, r, and α are positive constants;*
- *$b, d \in C_b^2\left(\mathbb{R}^d \times \mathbb{R}_+\right)$ are nonnegative functions;*
- *$\mathbf{c}: \mathbb{R}^d \times \mathbb{R} \to \mathbb{R}$ is bounded and boundedly differentiable;*
- *$\kappa \in C_b^2\left(\mathbb{R}^d\right)$ satisfies $\int_{\mathbb{R}^d} \kappa(\mathbf{x}) d\mathbf{x} = 1$.*

Then for each $T > 0$ there exists a unique solution of (7.13), (7.15) on $[0, T]$ and

(i) *the sequence of processes (ξ_t^N) given by (7.14) converges in distribution to $\bar{\mu}_t$ defined by (7.19) on $\mathbb{D}([0, T], \mathscr{M})$ with Skorohod topology. The space $\mathscr{M}$ is considered here with a topology of vague convergence;*
(ii) *the sequence of processes ρ_t^N converges to ρ_t given by (7.18) in distribution on $C([0, T], C_b^1(\mathbb{R}^d))$.*

The pair $(\bar{\mu}, \rho)$ coincides with $(\tilde{\mu}, \rho)$ and is the weak solution of (7.16).

The proof of this theorem was inspired by Budhiraja and Fan [20]. In the classical propagation-of-chaos-type convergence, one usually can prove also the pathwise convergence for a single particle. Here, a single particle lives for a random period of time, and either dies or has offspring, so such a convergence is not possible. But if we consider a whole cell line of a single cell (i.e. one initial cell with all its descendants) as an element of the space $\mathbb{X}$ (see Sect. 7.2) and introduce metrics on this space, then we can prove pathwise convergence of the whole cell line. We define metrics by

$$d_{\mathbb{X}}(\mathrm{x}, \mathrm{y}) = \sup_{(i,j) \in \mathbb{N} \times \mathbb{J}} \min\{|\mathbf{x}_{i,j} - \mathbf{y}_{i,j}|, 1\} \qquad (7.22)$$

for $\mathrm{x} = (\mathbf{x}_{i,j})_{(i,j) \in \mathbb{N} \times \mathbb{J}}$, $\mathrm{y} = (\mathbf{y}_{i,j})_{(i,j) \in \mathbb{N} \times \mathbb{J}} \in \mathbb{X}$, with a convention that $|\mathbf{x} - *| = 1$ for any $\mathbf{x} \in \mathbb{R}^d$. Then, we have the following fact.

Theorem 7.2 *Let assumptions of Theorem 7.1 be satisfied. Denote by $\mathrm{x}_1^N(t)$ the cell line of the $(1, \emptyset)$-th initial cell of the solution of (7.15) and denote by $\bar{\mathrm{x}}^N(t)$ the solution of (7.17). Then for each $T > 0$*

$$\sup_{t \in [0,T]} d_{\mathbb{X}}\left(\mathrm{x}_1^N(t), \bar{\mathrm{x}}(t)\right) \qquad (7.23)$$

converges to 0 in probability as $N \to \infty$.

7.4 Other IBMs

We now present an overview of selected biological models in which the limit passages from individual models to nonlinear transport equations or superprocesses are investigated. The research on these issues was inspired by attempts to describe the birth and death process with respect to the spatial structure [7, 33] and models of mutation and genetic drift [48, 84]. Appropriate limit passages result in deterministic processes or superprocesses e.g. the Dawson–Watanabe and the Fleming–Viot superprocesses. The mathematical properties of these models and their generalizations have been intensively studied e.g. in [37, 106, 111], and a more comprehensive literature on this topic is given in the review article [40].

Another group consists of models inspired by evolutionary ecology and describing the distribution of phenotypes in a population taking into account mutations and adaptive processes [27, 29, 62, 78]. These types of models can be studied quite accurately using limit passage theorems from individual models to transport equations or superprocesses. For example in [62] the limit passage leads to processes with Lévy fluctuations. In [27] it is considered a model describing the distribution of phenotypes in a population and, depending on the scaling used, different types of limits are obtained, both deterministic (ordinary and partial differential equations and differential-integral equations) and stochastic (superprocesses and SPDEs). The model considered in [78] includes the phenotypic and age structure of the population. The model investigated in [29] combines elements of Mendelian genetics with adaptive dynamics.

The third group are particle-based reaction-diffusion models. Reaction-diffusion systems of equations are commonly used to describe various phenomena in natural sciences such as chemistry and biology, e.g. pattern formation. They represent the situation where several (or sometimes only one) species diffuse in the space and undergo reactions, that can be chemical reactions or kind of birth and death process or transitions from one population to another. One of the first strict mathematical proofs of convergence of a stochastic interacting particles system to a reaction-diffusion system was given by Oelschläger [83]. Imagine a population consisting of $\approx N$ diffusing particles that are divided into several, say K, subpopulations. The particles undergo diffusion and randomly, possibly depending on the state of the whole population, change the subpopulation. N is the scaling parameter as in previous chapters and can be regarded as the initial number of particles, but the actual number of particles varies in time. Let $M(N, r, t)$ be the set of indices of all particles that belong to r-th population at time t. We define the empirical processes of each subpopulation by

$$\xi_{N,r}(t) = \frac{1}{N} \sum_{i \in M(N,r,t)} \delta_{X_i}$$

where X_i is the position of i-th particle and δ_x is a Dirac delta. Oelschläger [83] proved that under appropriate assumptions the Wasserstein distance (cf. (5.11)) from the empirical measures $\xi_{N,r}(t)$ to the solution of the corresponding reaction-diffusion system of PDEs goes to zero as the number of particles go to infinity. The most challenging issues were related to the fact of varying in time size of the population and to the range of interactions. Namely, in the particle model, the interactions cannot be local, since the probability of meeting of two particles is 0 (unless the dimension is 1), so the size of particle is usually represented by some interaction kernel. This kernel may be rescaled with the size of population and Oelschläger proposed the rescaling of the form $V_N(x) = \alpha_N^d V_1(\alpha_N x)$, where $\alpha_N \sim N^{\beta/d}$ with $\beta \in (0, 1)$ and V_1 is some base kernel. This rescaling was called "moderate" and is widely used since then, e.g. [18, 47].

For the last forty years numerous approaches have been used to model reacting diffusive particles on individual level and still are of great interest. For example [47] the empirical process of one-species particle reaction-diffusion system converges to the solution of the Fisher–Kolmogorov–Petrowskii–Piskunov equation. A new method of the proof is presented, which was inspired by some SPDE methods and gives the uniform convergence in the space variable. Another, rather general, particle-based stochastic reaction-diffusion model is proposed in [58] where the convergence to a coarse-grained mean-field deterministic partial integro-differential equation is proven. In [86] two-scale individual-based stochastic reaction-diffusion model is considered with localized and nonlocalized species. The limit passage leads

7.4 Other IBMs

to a multiscale piecewise-deterministic model, where PDE are coupled to a jump birth and death type process.

The next group consists of models describing aggregation behaviour or interactions between different populations. The starting point was the work presenting a mathematical description of aggregation processes in the animal world [53–55]. Limit passages for such processes were studied in [12, 80, 82]. A model investigated in [12] describes the behaviour of ants of the genus *Polyergus*, among which there are no workers other than soldiers. They invade other anthills from which they steal larvae and pupae that are used after rearing as workers. When invading other anthills, these ants exhibit complex aggregation behaviour. The paper presents an individual soldier movement model in which there are interactions of attraction and repulsion between individuals and Brownian motion. A limit passage results in an equation similar to (6.19). An individual-based model of daphnia population is considered in [80]. By choosing appropriate limit passages, a deterministic model describing the population structure with respect to age and size of individuals and algal concentration is derived. Limit passages have also been investigated in evolutionary game theory models [30, 65, Sec. 6.10]. The reader interested in a concise discussion of limit passages issues is referred to Sec. 6.3 of the book [22].

Another class of recent models, that were not mentioned in this book, considers diffusing individuals with finite (nonzero) volume. Assuming that particles are just points is very common and comfortable, but it is always a simplification. Although often this simplification is reasonable, sometimes it is important to model the fact that individuals are not zero-dimensional points, but have finite size. This assumption leads to various mathematical intricacies. One of the possible approaches is to write the Fokker-Planck equations for the distributions of positions of all "hard ball" particles on the appropriately constructed domain, in such a way that the positions of centers are in the distance at least of two radii (cf. [17]). Let Ω be a bounded domain in R^d. If the particles were points the space of positions of N particles $\mathbf{x} = (x_1, \ldots, x_N)$ would be Ω^N. For the particles that are hard spheres of radius $\varepsilon > 0$, the positions of centers of particles are located in the "hollow" configuration space $\Omega_\varepsilon^{[N]} = \Omega^N \setminus \mathscr{B}_\varepsilon$ where $\mathscr{B}_\varepsilon$ is the set of all illegal configurations (such that at least two balls overlap)

$$\mathscr{B}_\varepsilon = \left\{ \mathbf{x} \in \Omega^N : \|x_i - x_j\| \leq 2\varepsilon \right\}.$$

In the simplest case of identical independent (besides the collisions) particles, one can write the Fokker–Planck equation for the distribution $p(\mathbf{x}, t)$ of positions of all particles in $\Omega_\varepsilon^{[N]}$

$$\frac{\partial p}{\partial t} = \nabla \cdot [\nabla p - \mathbf{F}(\mathbf{x}) p]$$

with a reflective boundary condition

$$[\nabla p - \mathbf{F}(\mathbf{x})p] \cdot \mathbf{n} = 0 \text{ on } \partial\Omega_\varepsilon^{[N]},$$

where $\mathbf{n}$ denotes the unit outward normal vector. The model, although linear, is hardly treatable, due to very complicated boundary condition, but if the initial distribution of the particles is the same, then the particles are interchangeable and p is symmetric to permutations of the particles. So one should be interested in the marginal distribution of a single particle $p_1(x_1, t) = \int p(\mathbf{x}, t) dx_2 \cdots dx_n$. Using the asymptotic expansion as the radius of particles goes to 0, Bruna and Chapman [17] obtain an approximated equation for the marginal

$$\frac{\partial p_1}{\partial t} = \nabla \cdot [\nabla(p_1 + \alpha_d(N-1)\varepsilon^d p_1^2) - F(x_1)p_1],$$

where α_d is a dimension-dependent constant, e.g. $\alpha_2 = \pi/2$ and $\alpha_3 = 2\pi/3$. This approach was extended to various models e.g. the situation of different species [16], and recently for non-spherical shapes such as hard needles [19]. It can be also incorporated in much more complex cases such as propagation of gradient flow structures from microscopic to macroscopic model [18].

Individual models and limit passages can be described and studied using yet other probabilistic tools. Such tools include random measures describing the evolution of the spatial distribution of a discrete set of points [66, 102]. This approach is equivalent to the individual description studied here. It allows to study both limit passages to meso- and macroscopic models, and correlation in the distribution of individuals.

References

1. Adioui, M., Arino, O., El Saadi, N.: A non local model of phytoplankton aggregation. Nonlinear Anal. Real World Appl. **6**, 593–607 (2005)
2. Adler, R.: Superprocesses and plankton dynamics. In: Monte Carlo Simulation in Oceanography, Proceedings of the 'Aha Huliko'a Hawaiian Winter Workshop, pp. 121–128. University of Hawaii at Manoa, Honolulu (1997)
3. Almeida, C.R., de Abreu, F.V.: Dynamical instabilities lead to sympatric speciation. Evol. Ecol. Res. **5**, 739–757 (2003)
4. An, G., Mi, Q., Dutta-Moscato, J., Vodovotz, Y.: Agent-based models in translational systems biology. Syst. Biol. Med. **1**, 159–171 (2009)
5. Arino, O., Rudnicki, R.: Phytoplankton dynamics. C. R. Biol. **327**, 961–969 (2004)
6. Aubert, M., Chaplain, M.A.J., McDougall, S.R., Devlin, A., Mitchell, G.A.: A continuum mathematical model of the developing murine retinal vasculature. Bull. Math. Biol. **73**, 2430–2451 (2011)
7. Bailey, N.T.J.: Stochastic birth, death and migration processes for spatially distributed populations. Biometrika **55**, 189–198 (1968)
8. Barton, N.H., Etheridge, A.M., Véber, A.: The infinitesimal model: definition, derivation, and implications. Theor. Popul. Biol. **118**, 50–73 (2017)
9. Billingsley, P.: Probability and Measure. Wiley, New York (1986)
10. Billingsley, P.: Convergence of Probability Measures. Wiley, New York (1999)
11. Bobylev, A.V.: Exact solutions of the Boltzmann equation. Sov. Phys. Dokl. **20**, 822–824 (1976)
12. Boi, S., Capasso, V., Morale D.: Modeling the aggregative behavior of ants of the species *Polyergus rufescens*. Nonlinear Anal. Real World Appl. **1**, 163–176 (2000)
13. Bolker, B., Pacala, S.: Using moment equations to understand stochastically driven spatial pattern formation in ecological systems. Theor. Popul. Biol. **52**, 179–197 (1997)
14. Bonabeau, E.: Agent-based modeling: methods and techniques for simulating human systems. PNAS **99**, 7280–7287 (2002)
15. Bréhier, C.E.: A short introduction to Stochastic PDEs. https://hal.archives-ouvertes.fr/hal-00973887v2/document
16. Bruna, M., Chapman, S.: Diffusion of multiple species with excluded-volume effects. J. Chem. Phys. **137**, 204116 (2012)
17. Bruna, M., Chapman, S.: Excluded-volume effects in the diffusion of hard spheres. Phys. Rev. E **85**, 011103 (2012)

18. Bruna, M., Burger, M., Carrillo, J.A.: Coarse graining of a Fokker-Planck equation with excluded volume effects preserving the gradient flow structure. Eur, J. Appl. Math. **32**, 711–745 (2021)
19. Bruna, M., Chapman, S.J., Schmidtchen, M.: Derivation of a macroscopic model for Brownian hard needles. Proc. R. Soc. A Math. Phys. Eng. Sci. **479**, 20230076 (2023)
20. Budhiraja, A., Fan, W.-T.: Uniform in time interacting particle approximations for nonlinear equations of Patlak-Keller-Segel type. Electron. J. Probab. **22**, Paper No. 8, 37 (2017)
21. Bulmer, M.G.: The Mathematical Theory of Quantitative Genetics. Clarendon Press, Oxford (1980)
22. Capasso, V., Bakstein, D.: An Introduction to Continuous-Time Stochastic Processes. Birkhäuser, Boston (2005)
23. Capasso, V., Flandoli, F.: On the mean field approximation of a stochastic model of tumor-induced angiogenesis. Eur. J. Appl. Math. **30**, 619–658 (2019)
24. Capasso, V., Morale, D.: Asymptotic behavior of a system of stochastic particles subject to nonlocal interactions. Stoch. Anal. Appl. **27**, 574–603 (2009)
25. Capasso, V., Morale, D., Facchetti, G.: The role of stochasticity in a model of retinal angiogenesis. IMA J. Appl. Math. **77**, 729–747 (2012)
26. Capasso, V., Wieczorek, R.: A hybrid stochastic model of retinal angiogenesis. Math. Methods Appl. Sci. **43**, 10578–10592 (2020)
27. Champagnat, N., Ferriére, R., Méléard, S.: From individual stochastic processes to macroscopic models in adaptive evolution. Stoch. Models **24**, 2–44 (2008)
28. Clark, P.J., Evans, F.C.: Distance to nearest neighbor as a measure of spatial relationships in populations. Ecology **35**, 445–453 (1954)
29. Collet, P., Méléard, S., Metz, J.A.J.: A rigorous model study of the adaptive dynamics of Mendelian diploids. J. Math. Biol. **67**, 569–607 (2013)
30. Corradi, V., Sarin, R.: Continuous approximations of stochastic evolutionary game dynamics. J. Econ. Theory **94**, 163–191 (2000)
31. Da Prato, G., Zabczyk, J.: Stochastic equations in infinite dimensions. Encyclopedia of Mathematics and its Applications. Cambridge University Press, Cambridge (2014)
32. Davis, M.H.A.: Piecewise-deterministic Markov processes: a general class of nondiffusion stochastic models. J. R. Stat. Soc. Ser. B **46**, 353–388 (1984)
33. Dawson, D.A.: Stochastic evolution equations. Math. Biosci. **154**, 187–316 (1972)
34. Dawson, D.A.: Measure-valued Markov processes. In: École d'Été de Probabilités de Saint-Flour XXI—1991. Lecture Notes in Mathematics, vol. 1541, pp. 1–260. Springer, Berlin (1993)
35. DeAngelis, D.L., Gross, L.J. (Eds.): Individual-Based Models and Approaches in Ecology: Populations, Communities and Ecosystems. Chapman and Hall, New York (1992)
36. Doebeli, M., Blok, H.J., Leimar, O., Dieckmann, U.: Multimodal pattern formation in phenotype distributions of sexual populations. Proc. R. Soc. B **274**, 347–357 (2007)
37. Donnelly, P., Kurtz, T.G.: Particle representations for measure-valued population processes. Ann. Probab. **27**, 166–205 (1999)
38. Dynkin, E.B.: An introduction to branching measure-valued processes. CRM Monograph Series, vol. 6. American Mathematical Society, Providence (1994)
39. El Saadi, N., Bah, A.: An individual-based model for studying the aggregation behavior in phytoplankton. Ecol. Model. **204**, 193–212 (2007)
40. Engländer, J.: Branching diffusions, superdiffusions and random media. Probab. Surv. **4**, 303–364 (2007)
41. Etheridge, A.M.: An Introduction to Superprocesses. University Lecture Series, vol. 20. American Mathematical Society, Providence (2000)
42. Ethier, S.N., Kurtz, T.G.: Markov Processes. Characterization and Convergence. Wiley, New York (1986)
43. Fan, J.Y., Hamza, K., Jagers, P., Klebaner, F.C.: Limit theorems for multi-type general branching processes with population dependence. Adv. Appl. Probab. **52**, 1127–1163 (2020)

References

44. Ferland, R., Fernique, X., Giroux, G.: Compactness of the fluctuations associated with some generalized nonlinear Boltzmann equations. Can. J. Math. **44**, 1192–1205 (1992)
45. Finkelshtein, D., Kondratiev, Y., Kozitsky, Y., Kutoviy, O.: The statistical dynamics of a spatial logistic model and the related kinetic equation. Math. Models Methods Appl. Sci. **25**, 343–370 (2015)
46. Fisher, R.A.: On the correlation between relatives on the supposition of Mendelian inheritance. Trans. R. Soc. Edinburgh **52**, 399–433 (1918)
47. Flandoli, F., Leimbach, M., Olivera, C.: Uniform convergence of proliferating particles to the FKPP equation. J. Math. Anal. Appl. **473**, 27–52 (2019)
48. Fleming, W.H., Viot, M.: Some measure-valued Markov processes in population genetics theory. Ind. Univ. Math. J. **28**, 817–843 (1979)
49. Fournier, N., Méléard, S.: A microscopic probabilistic description of locally regulated population and macroscopic approximations. Ann. Appl. Probab. **14**, 1880–1919 (2004)
50. Gavrilets, S., Boake, C.R.B.: On the evolution of premating isolation after a founder event. Am. Nat. **152**, 706–716 (1998)
51. Gihman, I.I., Skorohod, A.V.: Stochastic differential equations. Ergebnisse der Mathematik und ihrer Grenzgebiete, Band 72. Springer, New York (1972)
52. Grimm, V., Railsback, S.F.: Individual-Based Modeling and Ecology. Princeton University Press, Princeton (2005)
53. Grünbaum, D.: Translating stochastic density-dependent individual behaviour with sensory constraints to an Eulerian model of animal swarming. J. Math. Biol. **33**, 139–161 (1994)
54. Grünbaum, D., Okubo, A.: Modelling social animal aggregations. In: Levin, S. (ed.), Frontiers of Theoretical Biology. Lectures Notes in Biomathematics, vol. 100, pp. 296–325. Springer, New York (1994)
55. Gueron, S., Levin, S.A., Rubenstein, D.I.: The dynamics of herds: from individuals to aggregations. J. Theor. Biol. **182**, 85–98 (1996)
56. Hairer, M.: An introduction to stochastic PDEs (2009). arXiv:0907.4178
57. Itô, K.: Distribution-valued processes arising from independent brownian motions. Math. Z. **182**, 17–34 (1983)
58. Isaacson, S.A., Ma, J., Spiliopoulos, K.: Mean field limits of particle-based stochastic reaction-diffusion models. SIAM J. Math. Anal. **54**(1), 453–511 (2022)
59. Jacod, J., Shiryaev, A.N.: Limit Theorems for Stochastic Processes. Grundlehren der Mathematischen Wissenschaften, vol. 288, Springer, Berlin (2003)
60. Jagers, P.: General branching processes as Markov fields. Stoch. Process. Appl. **32**, 183–212 (1989)
61. Jakubowski, A.: On the Skorohod topology. Ann. Inst. H. Poincaré B **22**, 263–285 (1986)
62. Jourdain, B., Méléard, S., Woyczynski, W.: Lévy flights in evolutionary ecology. J. Math. Biol. **65**, 677–707 (2012)
63. Kiełek, Z.: Asymptotic behaviour of solutions of the Tjon–Wu equation. Ann. Polon. Math. **52**, 109–118 (1990)
64. Kokko, H., Jennions, M.D., Brooks, R.: Unifying and testing models of sexual selection. Ann. Rev. Ecol. Evol. Syst. **37**, 43–66 (2006)
65. Kolokoltsov, V.N.: Markov Processes, Semigroups and Generators. Studies in Mathematics, vol. 38. De Gruyter, Berlin (2011)
66. Kondratiev, Y., Kozitsky, Y.: The evolution of states in a spatial population model. J. Dyn. Differ. Equ. **30**, 135–173 (2018)
67. Krook, M., Wu, T.T.: Exact solutions of the Boltzmann equation. Phys. Fluids **20**, 1589–1595 (1977)
68. Krylov, N.V.: On SPDE's and superdiffusions. Ann. Prob. **25**, 1789–1809 (1997)
69. Kuno, E.: Simple mathematical models to describe the rate of mating in insect populations. Res. Popul. Ecol. **20**, 50–60 (1978)
70. Lachowicz, M.: From microscopic to macroscopic descriptions of complex systems. C. R. Mecanique **331**, 733–738 (2003)

71. Lachowicz, M.: Microscopic, mesoscopic and macroscopic descriptions of complex systems. Prob. Eng. Mech. **26**, 54–60 (2011)
72. Lasota, A.: Asymptotic stability of some nonlinear Boltzmann-type equations. J. Math. Anal. Appl. **268**, 291–309 (2002)
73. Lasota, A., Traple, J.: An application of the Kantorovich-Rubinstein maximum principle in the theory of the Tjon–Wu equation. J. Differ. Equ. **159**, 578–596 (1999)
74. Law, R., Dieckmann, U.: Moment approximations of individual-based models. In: Dieckmann, U., Law, R., Metz J.A.J. (eds.) The Geometry of Ecological Interactions, pp. 252–270. Cambridge University Press, Cambridge (2002)
75. Lomnicki, A.: Population Ecology of Individuals. Princeton University Press, Princeton (1988)
76. McDougall, S.R., Watson, M.G., Devlin, A., Mitchell, G.A., Chaplain, M.A.J.: A hybrid discrete-continuum model of pattern prediction in the developing retinal vasculature. Bull. Math. Biol. **74**, 2272–2314 (2012)
77. Méléard, S.: Convergence of the fluctuations for interacting diffusions with jumps associated with Boltzmann equations. Stoch. Stoch. Rep. **63**, 195–225 (1998)
78. Méléard, S., Tran, V.C.: Slow and fast scales for superprocess limits of age-structured populations. Stoch. Proc. Appl. **122**, 250–276 (2012)
79. Metz, J.A.J., de Roos, A.M.: The role of physiologically structured population models within a general individual-based modeling perspective. In: DeAngelis, D.L., Gross L.J. (eds.) Individual-Based Models and Approaches in Ecology, pp. 88–111. Routledge, London (1992)
80. Metz, J.A.J., Tran, V.C.: Daphnias: from the individual based model to the large population equation. J. Math. Biol. **66**, 915–933 (2013)
81. Mitoma, I.: Tightness of probabilities on $C([0, 1]; \mathscr{S}')$ and $D([0, 1]; \mathscr{S}')$. Ann. Probab. **11**, 989–999 (1983)
82. Morale, D., Capasso, V., Oelschläger, K.: An interacting particle system modelling aggregation behavior: from individuals to populations. J. Math. Biol. **50**, 49–66 (2005)
83. Oelschläger, K.: On the derivation of reaction-diffusion equations as limit dynamics of systems of moderately interacting stochastic processes. Probab. Theory Relat. Fields **82**, 565–586 (1989)
84. Ohta, T., Kimura, M.: A model of mutation appropriate to estimate the number of electrophoretically detectable alleles in a finite population. Genet. Res. Camb. **22**, 201–204 (1973)
85. Perkins, E.: Dawson-Watanabe superprocesses and measure-valued diffusions. In: Lectures on Probability and Statistics (Saint Flour 1999). Lecture Notes in Mathematics, vol. 1781, pp. 125–324. Springer, Berlin (2002)
86. Popovic, L., Véber, A.: A spatial measure-valued model for chemical reaction networks in heterogeneous systems. Ann. Appl. Probab. **33**(5), 3706–3754 (2023)
87. Rachev, S.T.: Probability Metrics and the Stability of Stochastic Models. Willey, Chichester (1991)
88. Railsback, S.F., Grimm, V.: Agent-Based and Individual-Based Modeling: A Practical Introduction. Princeton University Press, Princeton (2019)
89. Rejniak, K.A., Anderson, A.R.A.: Hybrid models of tumor growth. Wiley Interdiscip. Rev. Syst. Biol. Med. **3**, 115–125 (2011)
90. Remenik, D.: Limit theorems for individual-based models in economics and finance. Stoch. Process. Appl. **119**, 2401–2435 (2009)
91. Rudnicki, R., Tyran-Kamińska, M.: Piecewise Deterministic Processes in Biological Models. SpringerBriefs in Applied Sciences and Technology, Mathematical Methods. Springer, Cham (2017)
92. Rudnicki, R., Wieczorek, R.: Fragmentation – coagulation models of phytoplankton. Bull. Pol. Acad. Sci. Math. **54**, 175–191 (2006)
93. Rudnicki, R., Wieczorek, R.: Phytoplankton dynamics: from the behaviour of cells to a transport equation. Math. Model. Nat. Phenom. **1**, 83–100 (2006)
94. Rudnicki, R., Wieczorek, R.: Mathematical models of phytoplankton dynamics. In: Russo R. (ed.) *Aquaculture I. Dynamic Biochemistry, Process Biotechnology and Molecular Biology*, vol. 2 (Special Issue 1), pp. 55–63 (2008)

95. Rudnicki, R., Wieczorek, R.: Does assortative mating lead to a polymorphic population? A toy model justification. Discrete Contin. Dyn. Syst. Ser. B **23**, 459–472 (2018)
96. Rudnicki, R., Zwoleński, P.: Model of phenotypic evolution in hermaphroditic populations. J. Math. Biol. **70**, 1295–1321 (2015)
97. Rudnicki, R., Tiuryn, J., Wójtowicz, D.: A model for the evolution of paralog families in genomes. J. Math. Biol. **53**, 759–770 (2006)
98. Schneider, K.A., Bürger, R.: Does competitive divergence occur if assortative mating is costly? J. Evol. Biol. **19**, 570–588 (2006)
99. Schneider, K.A., Peischl, S.: Evolution of assortative mating in a population expressing dominance. PLoS ONE **6**, e16821 (2011)
100. Schuss, Z.: Theory and Applications of Stochastic Processes. Wiley, New York (1980)
101. Skolakis, G., Adler, R.: Superprocesses over a stochastic flow. Ann. Appl. Probab. **11**, 488–543 (2001)
102. Streit, R.L.: Poisson Point Processes: Imaging, Tracking, and Sensing. Springer, New York (2010)
103. Tjon, J.A., Wu, T.T.: Numerical aspects of the approach to a Maxwellian distribution. Phys. Rev. A. **19**, 883–888 (1979)
104. Uchmański, J., Grimm, V.: Individual-based modelling in ecology: what makes the difference? Trends Ecol. Evol. **11**, 437–441 (1996)
105. Vallender, S.: Calculation of the Wasserstein distance between probability distributions on the line. Theor. Probab. Appl. **18**, 784–786 (1973)
106. Verzani, J., Adler, R.: IBM, SIBM and IBS. Ann. Probab. **28**, 462–477 (2000)
107. Villani, C.: Optimal Transport, Old and New. Grundlehren der Mathematischen Wissenschaften, vol. 338. Springer, Berlin (2008)
108. Von Smoluchowski, M.: Drei Vorträge über Diffusion, Brownsche Molekularbewegung und Koagulation von Kolloidteilchen. Phys. Z. **17**, 557–571, 585–599 (1916)
109. Walsh, J.B.: An introduction to stochastic partial differential equations. In: Hennequin, P.L. (ed.) Ecole d'été de probabilités de Saint Flour XIV – 1984. Lecture Notes in Mathematics, vol. 1180, pp. 265–439. Springer, Berlin (1986)
110. Wieczorek, R.: Hydrodynamic limit of a stochastic model of proliferating cells with chemotaxis. Kinet. Relat. Models **16**, 373–393 (2023)
111. Xiong, J.: Long-term behavior for superprocesses over a stochastic flow. Elect. Commun. Probab. **9**, 36–52 (2004)
112. Zwoleński, P.: Trait evolution in two-sex populations. Math. Mod. Nat. Phenom. **498**, 163–181 (2015)
113. Zwoleński, P.: Mathematical models of phenotypic evolution in sexual populations and their asymptotic properties. Ph.D. Dissertation, IM PAN (2016)

Index

A
Additive noise, 76
Agent-based model, 1
Age-structured population, 8
Assortative mating, 11

B
Bag, 3
Branching Brownian motion, 54

C
C_c^∞ space, 22, 97
Càdlàg modulus, 41
Clark–Evans index, 94
Coagulation-fragmentation process, 12
Continuous-time Markov chain, 5
Convergence in the sense of distribution, 41
Covariation measure, 58
Critical branching process, 94

D
d_{TV} total variation metric, 28
$\mathbb{D}([0, T], X)$, 40
Dawson–Watanabe superprocess, 54
Description
  eulerian, 2
  lagrangian, 2
Dirac measure, 3

E
Einstein formula for the diffusion coefficient, 92

F
Fleminga–Viot superprocess, 63

G
Gaussian martingale, 34
Genealogical superprocess, 61
Generalised Tjon–Wu equation, 78
Global attractor, 72

H
Historical superprocess, 61

I
Indywidual-based model, 1
Infinitesimal generator matrix, 4
Itô formula, 39

J
Jump
  moments, 5
  rate function, 4, 5

K
Kato theorem, 5
Kolmogorov matrix, 4
Kolmogorov matrix non-explosive, 5

L
L^2-martingale measure, 58
Lagrangian description, 2
Limit set, 76

M
Macroscopic model, 2
Markov process
    piecewise deterministic, 5
    pure-jump, 5, 9, 10
Martingale
    Gaussian, 34
    measure, 58
Mating
    assortative, 11
    semi-random, 10
McKendrick, 28
Mesoscopic model, 2
Metric total variation, 28
Microscopic model, 2
Model
    agent-based, 1
    distribution of genes in the genome, 13
    indywidual-based, 1
    Sharpa–Lotka–McKendrick, 28
    structured, 1
Multiplicative noise, 76
Multiset, 3

N
Non-explosive Kolmogorov matrix, 5

O
Orthogonal martingale measure, 58

P
Paralogs, 13
Phenotype-structured population, 5, 9, 10
Phytoplankton, 12
Piecewise deterministic Markov process, 5
Point process, 2
Predictable σ-algebra, 59

Preference function, 79
Process
    coagulation-fragmentation, 12
Pure-jump Markov process, 5, 9, 10

R
Random measure, 19, 38
    compensated Poisson, 39
    integer-valued, 38
    Poisson, 38

S
Semi-random mating, 10
σ-finite measure with value in L^2, 58
Size-structured population, 7
Skorohod
    metric, 40
    topology J_1, 40
Space C_c^∞, 22, 97
Stochastic integral w.r. martingale measure, 59
Strategy, 1
Structured
    model, 1
    population w.r.t.
        age, 8
        phenotype, 9, 10
        size, 7
Superprocess, 18, 53
    Dawson–Watanabe, 54
    Fleming–Viot, 63
    genealogical, 61
    historical, 61

T
Tjon–Wu equation, 78
    generalised, 78
Total variation metric, 28
Trait, 2
Transient state, 13
Transition
    probability, 4
    probability function, 5
    rate, 4
    rate matrix, 4

W
Wasserstein distance, 73
White noise, 57